THE
PLANT
PROPAGATOR'S
BIBLE

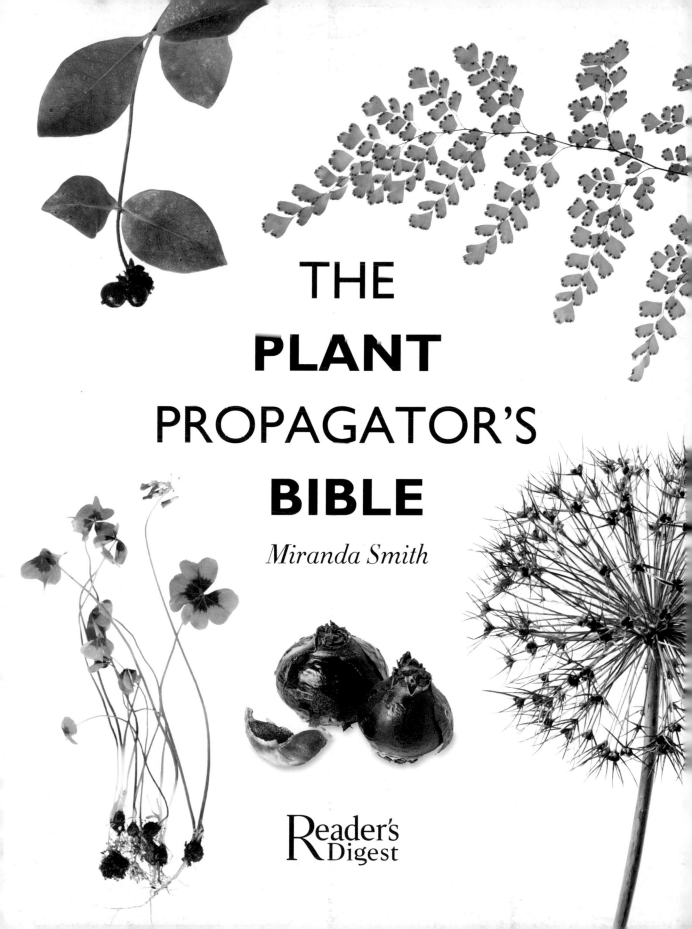

THE
PLANT
PROPAGATOR'S
BIBLE

Miranda Smith

Reader's
Digest

CONTENTS

A READER'S DIGEST BOOK

This edition published by
The Reader's Digest Association, Ltd.
11 Westferry Circus,
Canary Wharf,
London E14 4HE
United Kingdom

Conceived, designed, and produced by
Quarto Publishing plc
The Old Brewery
6 Blundell Street
London N7 9BH

QUAR.NPB

PROJECT EDITORS: Susie May, Karen Koll
ART EDITOR AND DESIGNER: Sheila Volpe
COPY EDITOR: Sue Viccars
ILLUSTRATOR: John Woodcock
PHOTOGRAPHERS: Paul Forrester,
Tim Himsel
PROOFREADER: Diana Chambers
INDEXER: Dorothy Frame
CONSULTANT: Catherine Gamble
PHOTO RESEARCHER: Claudia Tate
ASSISTANT ART DIRECTOR: Penny Cobb

ART DIRECTOR: Moira Clinch
PUBLISHER: Paul Carslake

ISBN 13: 978-0-276-44413-5

For more Reader's Digest products and
information visit our website at
www.readersdigest.co.uk

Manufactured by Modern Age Repro
House Ltd., Hong Kong
Printed by Star Standard
Industries (PTE) Ltd., Singapore

INTRODUCTION

Plant propagation is one of the most enjoyable – and certainly one of the most exciting – parts of gardening. Once gardeners are comfortable with building soils, keeping pests and diseases under control, and managing weeds without a lot of time or trouble, they naturally gravitate to experimenting with propagation – at least, most of them do. Unfortunately, some very good gardeners are frightened to take this step because they are convinced that they will "fail".

The sad thing about this decision is that it's based on very faulty information: the idea that plant propagation is difficult. Nothing could be further from the truth. After all, what is a plant designed to do? As far as the natural world is concerned, the only purpose of a plant is to make more plants. So you have the single most powerful ally in the world – Mother Nature – on your side when you try to propagate a plant.

You do have to follow some commonsense guidelines if you want to be successful. Whenever possible, you'll need to emulate natural processes as closely as you can. Every plant is designed to reproduce in certain ways – not all plants develop suckers, for example, nor can all of them make roots from a piece of leaf. So you'll need to learn how your plant reproduces in order to succeed.

That's the purpose of this book – to teach you how to propagate any plant in your garden.

The first section shows, in a step-by-step fashion, how to carry out various propagation techniques. You'll see that some require you to know a little about a plant's physiology, so that information is included as necessary. You'll find a list of the plants for which the technique is appropriate, and the index, beginning on page 189, also lists each of these plants so you'll be able to quickly find the information you need.

The second half of the book is devoted to a plant directory that covers plants in-depth. The plants were chosen because they are representative of the most common plants in contemporary gardens and include some trees, bushes and fruiting plants as well as ornamentals.

Most plants reproduce through a number of different processes. A plant may make seeds, for example, in addition to rooting easily from any branches that touch the ground, or from stem cuttings. In those cases, most people want to know which technique will give them the greatest chance of success. You'll find that information in the directory. Additional methods are also listed. This information is useful if you've missed your chance for the year to propagate the plant by the easiest means. For example, starting from seed is frequently the best way. However, if you suddenly decide to propagate something and it's too late in the year to buy seeds, it's useful to know that the plant also propagates well from semiripe cuttings, which are usually ready in mid- to late summer, or hardwood cuttings, which are ready in late autumn and early winter.

FAILURE AND SUCCESS

Even though most of your forays into the world of propagating plants will succeed, you won't have a 100 per cent success rate once you venture past working with seeds, suckers, divisions and layering. But it's important not to get discouraged. No one has a perfect record; even professionals experience some "failures".

Do yourself a favour by learning not to be upset if a graft doesn't take or a cutting develops a fungal disease. Instead, stop and analyse the situation. On the graft, for example, ask questions about your technique. Think back to try to determine if the cambium layers – which you'll learn about in this book – were really touching each other. Was the tape tight enough? Should you have sealed it with wax to exclude air and moisture?

Asking yourself these questions will generally lead to the answers that will result in success the next time around. And if it doesn't? Just ask again and try again. Unless you are trying to do something that's impossible – make an avocado leaf root, for example – you'll eventually succeed. And all those "failures" will be the reason you did – because you'll learn as much or more from your failures as you will from your successes.

HOW TO USE THIS BOOK

The Tools and Techniques section provides general advice for getting started and specific instructions for dozens of propagation techniques. The Plant Directory section discusses specific plants, with tips tailored to each genus.

List of plants appropriate for the featured technique

Checklist of tools and conditions

Tools and Techniques

Dividing plants with rhizomes

When you think of rhizomes, you probably think of the bearded iris with its prominent rhizomes emerging from the ground. But as you can see from the list at right, this certainly isn't the only garden plant that grows and can be propagated from a rhizome.

Timing matters when you are dividing rhizomes. In most regions it's best to wait until after the plant has finished blooming to dig up the rhizome and divide it. But if you live in a cold, short-season area, this may be impractical because divisions planted in late summer and early autumn don't have time to become established before the ground freezes. In this case, divide in spring and snap off any flower buds that form so the plant can put its energy into creating a strong root system for a year before it blooms.

You may wonder how often to divide rhizomatous plants. Some of them suffer if their rhizomes are allowed to grow unchecked, so dividing them every 3 to 5 years keeps them vital and strong. Once again, rather than adhering to a predetermined schedule, watch the plant and divide it when flowering decreases, flower size diminishes, or it just looks crowded.

APPROPRIATE PLANTS

COMMON	BOTANICAL
Cupid's Bowers	Achimenes spp.
Baneberries	Actaea spp.
Maidenhair Ferns	Adiantum spp.
Elephant's Ears	Alocasia spp.
Peruvian Lilies	Alstroemeria spp.
Wild Gingers	Asarum spp.
Blackberry Lily	Belamcanda chinensis
Bergenias	Bergenia spp.
Cuckoo Flower	Cardamine pratensis
Bearded Irises	Iris spp.
Lotus	Nelumbo spp.
Mayapples	Podophyllum spp.
Tuberose	Polianthes tuberosa
Solomon's Seals	Polygonatum spp.
Wall Polypody, Rockcap Ferns	Polypodium spp.
Christmas Ferns, Sword Ferns	Polystichum spp.
Bloodroot	Sanguinaria canadensis
Wakerobins, Wood Lilies	Trillium spp.

Detail photos showing relevant close-up views

rhizome

Colour photos displaying the featured propagation technique

stems and leaves

roots

What can go wrong
Rhizome doesn't grow: if you plant a rhizome with no buds, it cannot grow leaves and will die. If you are cutting a rhizome before the leaves have emerged, check to make certain that each piece has at least one, but preferably two, strong growth buds.

This rhizome should have more growth buds.

LIFT
Dig up the entire rhizome of the plant you want to divide, trying not to damage the root system.

EXAMINE
Divisions will reestablish best if they have at least two growth buds with leaves growing from them as well as many roots.

CUT
Use a very sharp knife to cut the divisions. A blunt knife can bruise tissues because you have to press so hard on it.

Checklist
Season: Spring or after plants finish blooming

Tools: Garden fork, trowel, garden knife, pruners

Equipment: Garden hose, cutting board

Supplies: Compost, soil improver

Temperature: Cool, not windy

Humidity: Not important

Light: Cloudy

TRIM
Cut back the leaves so that the plant isn't stressed by having to provide water for a large leaf area.

PLANT
Replant the rhizome at the same depth it was growing, and water it well to exclude air pockets from around the roots.

Propagate Bergenia by dividing rhizomes.

Discussion of potential problems with accompanying photo

Photo to support instructive text

Step-by-step illustrations and text demonstrating propagation technique

Plant Directory

Genus, common and family names, confirmed with Royal Horticultural Society references

Icons showing appropriate propagation techniques and hardiness zones

---

PROPAGATION METHODS

Easiest: Seed. Stratify the seeds for at least 2 weeks in the refrigerator before planting in a humus-rich soil mix. Seeds require dark for germination; after lightly covering them with soil, put a thick layer of newspapers or cardboard on top of the tray. Place the tray where the soil will remain at temperatures of 10° to 13°C. Seeds germinate within 2 weeks.

Additional methods: Division. Divide plants in early autumn, and replant immediately.

Basal cuttings. It's possible to root young shoots emerging from the crown. Take them when they are 10 to 15cm long, and root them in a propagating chamber that maintains temperatures of 16° to 18°C.

Potential Problems

Seeds go dormant if subjected to high temperatures, so it's important to hold the seed trays where they simply can't overheat. This is made easier by the seeds' requirement for darkness – they do well in a cool basement rather than a hot sunny room. Cuttings are prone to fungal infection. Use fresh, stem rot, less media to minimize problems and keep air circulation high.

Divisions will fail if they do not become established by the time the ground freezes. Divide and replant at least 40 days before your first expected autumn frost.

Dianthus spp.
Pinks, Carnations
CARYOPHYLLACEAE

Zones: 3–9

The 300 species in this genus come from Europe and Asia. They are grown for both their foliage, which in species such as maiden pinks (D. deltoides) forms a dense green mat of slender leaves, and their flowers, which range from the tall, perpetually flowering carnations found in commercial bouquets to the lovely sweet William (D. barbatus) blooms that get the summer border off to a colourful start. This diverse genus includes annuals, biennials and perennials, and flowers can be fragrant or not. Flower colours are white, pink, red and yellow, but the striking features are the stripes, picotees and target patterns on the petals.

PROPAGATION METHODS

Easiest: Seed. Seed for all types of pinks and carnations is widely available. Plant inside in early spring, and keep seed trays at temperatures of 16° to 21°C. Seeds will germinate in 2 to 3 weeks.

Additional methods: Layering. Tall perennial plants can be layered. As soon as you can bend a stem without breaking it, make a small wound where it will come into contact with the soil, and pin it to the soil surface. If practical, layer it in a pot sitting next to the plant because you won't have to bend the stem so close to the ground. It will root within a month. Sever the stem from the parent, and grow it on in the pot until early autumn, when you can transplant it to the garden.

Potential Problems

Seeds are close to foolproof, but seedlings of some species are quite tiny. Consequently, they are easy to kill with poor watering practices; mist them with a hand-mister held far enough away so that they never feel a blast of air or water. Layered plants will die over the winter if they are not well enough established to withstand the freezing and thawing action that is normal in most locations. If you don't think the plant has had time to get established in time for winter, sink the whole pot in the soil. In spring, you can dig it up and transplant the young plant to the garden.

Dicentra spp.
Bleeding Hearts, Dutchman's Breeches
FUMARIACEAE

Zones: 3–9

The more than 20 species in this genus come from Asia and North America. They are grown as much for their delicate foliage and bushy plant habit as for their long panicles of heart-shaped flowers. Flower colours include white, yellow and pink, and blooms are often bicoloured. Plants range in size from 36cm to 1.5m tall, but the most commonly seen species, such as fringed bleeding heart (D. eximia), are about 60cm tall and wide.

PROPAGATION METHODS

Easiest: Seed. Seed for many species of bleeding heart is widely available. Stratify it for 2 to 3 months before planting it on the surface of the seed tray. Fill the tray with a compost-based mix, and add a layer of vermiculite to cool temperatures. The seeds need light to germinate and may take as long as 3 months to do so. Consequently, it's best to cover the tray with plastic film and set it where the soil temperature will remain around 13° to 16°C.

Additional methods: Division. Divide plants in early spring, before new growth starts, and replant immediately.

Potential Problems

Seeds will not germinate if they do not experience a prolonged chilling period or if they overheat while they are in the seed tray. However, if you satisfy both these conditions, germination is excellent. The root system of bleeding heart is large enough to supply the top growth while the plant is getting established. Don't divide them into pieces that are too small, and keep them well watered after you have replanted them.

Echinacea spp.

E

Echinacea spp.
Coneflowers
ASTERACEAE

Zones: 3–9

The nine species in this genus come from North America. They are grown primarily for their flowers, which bloom in shades of white, pink and purple. The roots of purple coneflower (E. purpurea) are said to have medicinal qualities, so some people grow the plant to make tinctures. Favourite garden cultivars include 'White Swan', with white flowers and deep orange centres, and 'Magnus', with huge deep purple flowers with dark orange centres.

PROPAGATION METHODS

Easiest: Seed. Start seeds early inside, and keep the seed tray at about 13° to 16°C. They will germinate in 2 to 3 weeks.

Additional methods: Division. These plants do not tolerate a lot of root disturbance, so it's best to divide them as soon as the ground has thawed enough to dig. Rather than lifting the entire root ball, expose it on one side. Cut through the root ball, and working carefully, lift only that portion of the plant. Immediately fill in the hole with soil and replant the division. Mulch it well to retain moisture over the first season.

Root cuttings. You can take root cuttings in autumn. Do this without digging the plant. Again, simply expose a part of the root ball. Lay the root cuttings horizontally in the rooting medium rather than standing them upright, and cover them with about 0.5cm of medium. Place the rooting medium in a cool place in the house. They will root and send up shoots by spring. Alternatively, you can bury the cuttings in the soil under a cold frame to overwinter outside.

Potential Problems

Seeds will not germinate well if overheated; maintain cool temperatures for the best results. As explained above, divisions can be tricky. However, if you take proper precautions, you won't have trouble. Root cuttings are as susceptible to fungal infections as any other cuttings. Make certain that the medium is sterile when you place the cuttings in it, and cover it tightly with plastic wrap as soon as they are buried in it.

Eranthis spp.
Winter Aconites
RANUNCULACEAE

Zones: 4–9

The seven species in this genus come from Europe, Asia and the Far East. They are grown for their lovely little flowers that are often the first blooms you see in spring. The flowers of most species and cultivars are a clear, bright yellow, but E. pinnatifida, an alpine plant, has white flowers. The plants naturally form large masses if the leaves are left uncut until they wither. They prefer growing in a semishaded, damp area with alkaline to only slightly acid soil.

PROPAGATION METHODS

Easiest: Division. Tubers naturally increase, and plantings will enlarge every year. To separate the plants and give them more room or move some to another spot in the garden, separate and transplant the tubers just after the plants bloom – don't wait until autumn, as you do with many springtime bulbs.

Additional methods: Seed. Speciality seed companies sell seed for winter aconite, and you can also collect your own. As soon as the seed is ripe, plant it on the soil surface in a small tray, enclose the tray in plastic wrap and put it in the refrigerator. Leave it there for at least 3 weeks to 1 month. Then take it out and put the tray in a cold frame. Some seeds may germinate right away; some may wait until the following spring. Prick out seedlings after they have their first true leaves, but do not disturb the rest of the tray. Grow seedlings in containers sunk into the soil under the cold frame if plants are not large enough to transplant by autumn.

Potential Problems

Winter aconite does not fare well if its tubers dry out at any time. When buying tubers, buy only those that are packed in moist peat, otherwise they won't sprout. Remember this when making your divisions, too. Quickly move them – roots, tuber, shoots and all – to moist soil when you divide. Seed germination is really erratic; don't count a tray "dead" for at least 18 months after planting it.

Eryngium spp.
Sea Hollies
APIACEAE

Zones: 3–9

The more than 200 species in this genus come from Europe, North Africa, Turkey, Asia and Korea. They are grown for their unusual conelike, dry, blue-green, lavender-green, or grey-green flowers with prominent surrounding bracts. The blooms are long lasting in the garden and make a strong statement in any border. However, they can make an even stronger statement in

Eryngium spp.

Plant details and propagation advice, with species information confirmed with Royal Horticultural Society and American Horticultural Society references

Photo of a directory plant

TOOLS AND TECHNIQUES

In nature, plants propagate themselves without using special tools or sophisticated techniques. If branches from adjacent trees of the same genus rub against each other, for example, they may form a graft; the stem tips of leggy shrubs often take root; and seeds of plants as diverse as marigolds and giant sequoias germinate and grow without any fuss. But if you want to propagate the plants you want when you want them, you'll need a few basic tools and commonsense techniques to ease your work.

THE PROPAGATOR'S TOOL KIT

Good tools make an enormous difference to the ease with which you'll be able to propagate your plants. They can also make the difference between failure and success. So your first task as a plant propagator is to collect the tools you'll need for the jobs you'll be doing.

When you see all the tools together, it may appear as though your tool kit will cost a great deal to assemble. However, that isn't so. For one thing, unless you practise every type of propagation technique in these pages, you won't need every tool. Instead, you'll need to assemble certain tools for particular tasks. And as for cost, most of the tools are inexpensive, especially compared with the price of new plants. So regard these pages as a reference to come back to as you add to your repertoire of propagation techniques.

Health and safety

Gardening is good for you. Not only is it good exercise, it can provide real relaxation and enjoyment, both of which are essential to sound mental health. However, it can also pose dangers, not only to you, but also to your children and pets.

You'll use a variety of sharp knives and pruning tools when you propagate. As always, the first rule of working with cutting instruments is to keep them sharp. If you have to strain or push to cut with a dull blade, it's more likely to slip. You run the risk of cutting yourself and also injuring the plant. Sharpen the blades every time you use a gardening tool and you'll never have to worry about this.

Even if you use absolutely no synthetic chemicals in the garden, your success as a propagator sometimes depends on rooting hormones; strong disinfectants; and fungicides such as copper, sulphur, and Bordeaux mix. You may even buy one sort of fungus to protect from other fungi that cause diseases. Store these materials where children and pets can't get into them, and use protective gear (see below).

rubber gloves

Safety supplies

Use the protective gear listed here to prevent accidents when you are propagating. Remember to keep the first-aid kit well supplied.

Protective face mask
Plastic goggles
Rubber gloves
First-aid kit, including:
• Bandages
• Antibiotic ointment
• Gauze pads
• Iodine or alcohol prep pads

• Butterfly stitches
• Medical adhesive tape
• Scissors
• Support bandages with clips
• Aspirin and/or nonaspirin pain reliever

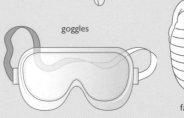

goggles

face mask

first-aid kit

Hand tools

These are the mainstay of the propagator's tool kit. You'll rely on pruners and knives to take cuttings, knives to make grafts, and spades and spading forks to divide plants. While you needn't break the bank to equip yourself, it is best to buy the highest-quality tools you can afford because they last the longest and generally do the best job.

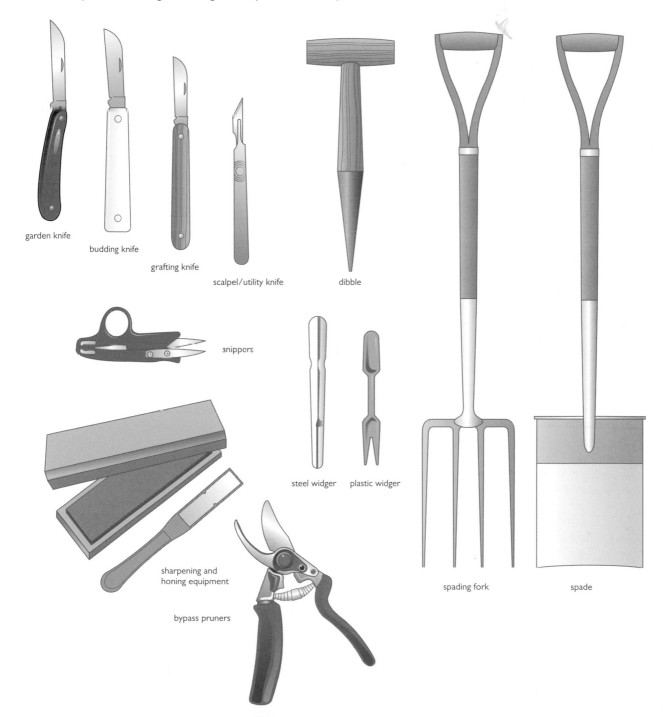

garden knife

budding knife

grafting knife

scalpel/utility knife

dibble

snippers

steel widger   plastic widger

sharpening and
honing equipment

bypass pruners

spading fork

spade

Equipment and supplies

Many of the items you'll use for propagation are things you already have in your gardening tool kit. You'll have to buy other items, such as grafting tape and rooting hormone. But no matter whether your equipment is new or old, set aside a special place to store all your propagation supplies.

grafting tape

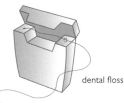

dental floss

twine

budding tape

grafting wax

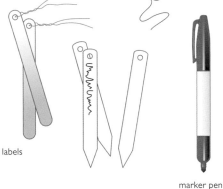

labels

marker pen

disinfectant

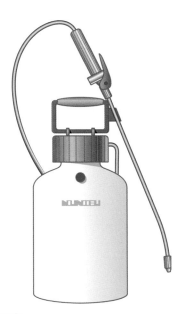

manual compressed air sprayer

pest management systems

Tool maintenance

A little care goes a long way when it comes to maintaining your tools in mint condition. As mentioned on page 12, sharpen all knives and even the edge of your spade before every use. After using, wipe them with a damp cloth to remove soil and sap. To keep diseases down, follow this with a spray of disinfectant, and after they dry, use an oily cloth to thinly coat all metal and wood parts. To keep your spade and spading fork in like-new condition, fill a 19-litre bucket with sand and add some used motor oil. Stir the oil into the sand. After every use, plunge the spade and spading fork up and down in the oily sand. Not only will this clean the tools, it will also leave them with a protective coating.

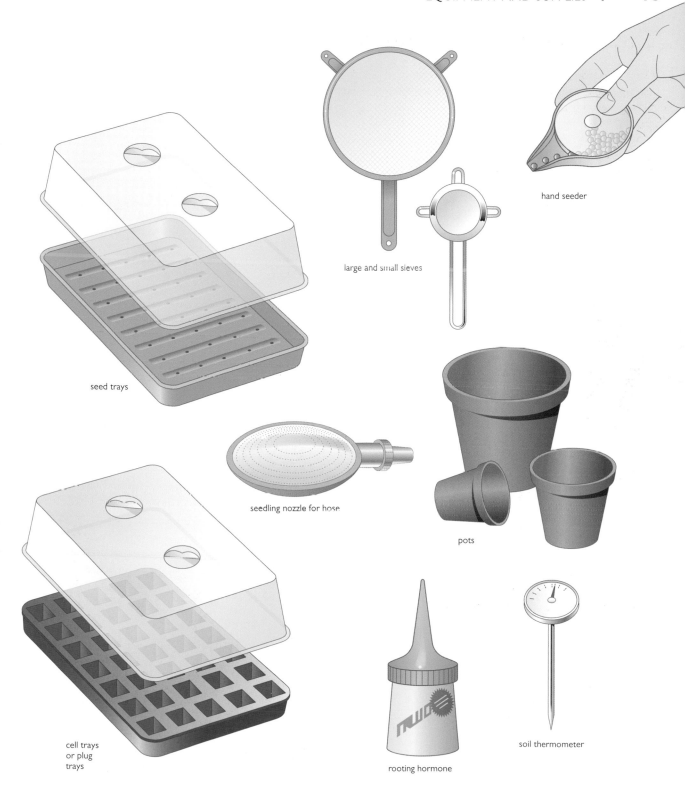

hand seeder

large and small sieves

seed trays

seedling nozzle for hose

pots

cell trays
or plug
trays

rooting hormone

soil thermometer

Soil mixes

You'll need a variety of soil mixes and media to propagate your plants. Some seeds are so prone to soilborne damping-off diseases that they germinate best in soil-less mix, and similarly, many types of cuttings require a sterile environment in which to root. In contrast, other seeds need nutrition so quickly that you get your best results by starting them in a nutrient-rich mix. Throughout this book, directions will include suggestions for the proper soil mix or medium for the plants and techniques.

Growing cuttings or seeds in cell trays reduces the risk of disease contamination.

Soil mix ingredients

Coarse sand: Use coarse washed sand to radically increase drainage and aeration in a mix. It is most useful in mixes for rooting cacti and other desert plants.

Coir: Made from coconut hulls, coir is a good substitute for medium-milled peat moss in soil-less media.

Compost: Made from decomposed organic matter, this material has a high cation exchange capacity, contributes balanced macro- and micronutrients and contains large populations of beneficial microorganisms that help protect plants against diseases.

Fine sand: Use fine washed sand to increase drainage and aeration in a mix. Because they're irregularly shaped, inert particles allow water to pass quickly and create voids for supplies of air.

Leaf mould: Decomposed leaves have a high cation exchange capacity and good mineral content but need macronutrients. Leaf mould often has a low pH and maybe pests and diseases.

Peat moss: Use finely milled peat moss in germinating mixes, and save the medium-milled material for sterile media for rooting cuttings. Do not use coarsely milled material for propagation.

Perlite: Made from expanded volcanic rock, perlite holds water in its cavities; because it is inert, it also sheds water easily. It is commonly used in sterile mixes to add aeration and drainage.

Topsoil: Purchase only the highest-quality topsoil for use in propagation mixes. Ideally, it should be free of pest eggs, weed seeds and disease organisms.

Vermiculite: Made from expanded mica, a form of clay, this material contains trace amounts of magnesium and has a high cation exchange capacity. However, it still aids drainage and aeration.

Mini glossary

Cation exchange capacity: The capacity to hold positively charged nutrient ions, or cations, such as calcium, potassium and magnesium.

Inert: Material without a cation exchange capacity or the ability to contribute nutrients to a medium.

Macronutrients: Nitrogen, phosphorus and potassium in forms that are available to plants are generally considered macronutrients; some people also include calcium, magnesium and sulphur in this group.

Micronutrients: Nutrients that plants need in trace amounts, such as copper, boron, iron, manganese and zinc, are considered micronutrients.

pH: The measure of a medium's acidity and alkalinity on a scale from 1 to 14, where 7 is neutral. Numbers lower than 7 indicate acidity, and those above 7 indicate alkalinity.

Soil-less mix: A mix made from materials such as perlite, vermiculite, sand and peat moss.

Sterile mix: A soil mix made from inert substances or materials that have been sterilized. These mixes do not remain sterile for long but offer some protection nonetheless.

Soil mix recipes

When you make your own soil mixes, you can be certain what they contain and adjust them as necessary for particular uses. The following recipes are guidelines; adjust them to suit your materials and create specialized mixes.

NUTRIENT-RICH MIX
2 parts compost
1 part topsoil
1 part sand
1 part vermiculite
1 part perlite

Adjust the ratio of these ingredients to create a quickly draining medium that contains enough compost and topsoil to retain the "soil" smell. Test drainage by filling a 30cm-deep pot with the mix and watering it until moisture runs out of the drainage holes in the bottom of the pot. If water puddles on the soil surface or the mix is slow to drain, add more sand or perlite.

A good medium (above) produces good plants (below).

SOIL-LESS MIX
1 part perlite
1 part vermiculite
2 parts finely milled peat moss

Wearing a facial mask and goggles, cut a hole in the bag of perlite and pour warm water into it before scooping out to use; the water will help to prevent small, sharp particles from flying into your face, eyes and nose. Warm water also penetrates peat moss better than cold, so you'll need to moisten all ingredients with very warm water and keep them from drying out at any time.

STERILE MIX
1 part finely milled peat moss or coir
1 part perlite

If you need an extra-fine medium for seed starting, sieve the peat moss or coir through a kitchen sieve or screening material. Wearing a facial mask and goggles, treat perlite as directed above, and use very warm water to mix ingredients. Do not let dry.

Propagation equipment

You won't need every item on these pages when you begin propagating, but eventually, you'll probably want to have them all. Again, add to your collection as time and money allow to avoid overwhelming your budget.

Artificial lights

Inside, you'll need artificial lights to grow most of your seedlings and rooted cuttings. Choose from fluorescent tubes or specialized grow lights, and follow directions on the following pages for the length of time to use them and to determine desirable light intensities.

Soil heating mats

Many seeds require warm soil to germinate. If you are starting them inside the house, chances are that ambient temperatures won't create these conditions. Use a heating mat under your seed trays, and your seeds will germinate easily and quickly. Choose a mat with a thermostat if you start many seeds this way; otherwise, a simpler design will still do the trick.

Propagation units

Although you can get away with nothing more than a plastic bag supported by canes or wires over a pot when you are rooting easily established cuttings, you'll have greater success with more difficult plants if you use an electric propagation chamber. Choose between those that mist at regular intervals and those that create fog. Both supply adequate levels of humidity.

Nursery bed

Set aside a deep growing bed in your garden for starting seeds or holding plants that you'll later set out in your garden proper. If space in this area runs tight – as it might if you begin to grow a lot of biennials or young woody plants – find another area to devote to

Cold frames make ideal environments for propagating many plants, both during the growing season and in winter.

it. Ideally, part of the nursery bed will be sited in full sun and part will be in filtered shade for a few hours a day.

Cold frames

Cold frames are ideal spots to start some types of seeds and cuttings, and they are also useful for overwintering cuttings or exposing seeds to the periods of cold weather that will stimulate their germination in the spring.

If you aren't always home during the day, choose a cold frame with vents that open and close automatically as temperatures rise and fall. This simple feature will keep the cold frame cool enough so your seedlings or cuttings don't fry in the heat, and warm enough so they don't suffer from the cold.

Site the cold frame where it gets filtered light. Once it's time to harden off seedlings and get them adjusted to full sun and bright light, you can move their trays to an open area in the garden. Until then, both they and any cuttings you are starting will appreciate filtered light.

Mini "greenhouses"

A real greenhouse is a propagator's delight. However, not everyone has room for one in the budget or the garden. If this is your situation, choose a version designed for a windowsill or to sit on a porch or balcony.

Controlling the environment is essential in any greenhouse – big or small. Keep a min-max thermometer in it to monitor temperatures so you can open vents or the door as necessary to maintain desirable temperatures. Humidity must also be kept high enough for your plants' good health, but not so high that it stimulates fungal diseases. If the area dries out too quickly, set a bowl of gravel in the bottom of the greenhouse and fill it with water. As the water evaporates, humidity levels will increase. If humidity levels are too high, the solution is simple: just open the vents or a door to let the moist air flow out and drier air flow in. If humidity is impossible because ambient humidity levels are also too high, set an oscillating fan so it blows air across the plants in the miniature greenhouse.

Greenhouse

If you're lucky enough to own a greenhouse, set aside a bench or two for propagation. Depending on the plants you're working with, you can install a frame on which to hang shade cloth or burlap to create light shade, and take it off when you want the area to see full light. You'll have to control the greenhouse environment carefully to provide appropriate light, temperature and CO_2 conditions for all your plants, but once you can do that, it will be a joy to propagate in your greenhouse.

Above and right: Mini greenhouses are an affordable option.

Below: Many gardeners dream of having a small greenhouse.

STARTING FROM SEED

Seeds are true wonders. Inside the seed coat, a tiny plant, complete with its radical, or first root; stem; and cotyledons, or seed leaves, lies ready to take in water and begin growing. As you'll see on the following pages, your job as a propagator is to provide the appropriate amounts of water, light and oxygen at the correct temperatures to stimulate this tiny plant to burst through its seed coat and begin its new life.

Starting plants from seed may seem like the most mundane propagation method possible. However, once you branch out from tried-and-true garden annuals such as marigolds and lettuce, you'll discover an exciting world that's anything but routine. With the help of a few specialized techniques, you'll soon be growing trees, shrubs and exotic ornamentals that would be prohibitively expensive to buy as plants. As

well, you may begin breeding strains of vegetable and flower cultivars that are particularly well adapted to your environment, and you can learn how to breed back from hybrids to create open-pollinated plants that come true from your own collected seed. All it takes is a little know-how, some seeds and a few pieces of rudimentary equipment.

Top left: Achillea seeds need light to germinate. Center left: Acanthus seeds are protected by spines. Bottom left: It's easy to save seed from open-pollinated Papavers.

Many seeds aren't ripe until they are really dry and ready to fall out of the seedhead. Let plants dry well before enclosing the seedheads in paper bags or muslin bags to catch falling seeds.

Seed characteristics

Seeds come in various sizes and shapes—all designed to help the seed travel to a place where it has a chance to germinate and survive, or where it will be protected from its enemies. Some seeds have wings to help them fly to new locations; others have hooks that grab on to the fur of passing animals and clothing. Some come with tiny floss parachutes to catch the wind, and others come in buoyant pods or hulls that carry them down a river or across an ocean.

The Acer has winged seeds.

Seeds with wings or lightweight pods: Wings enable seeds to catch wind currents that will take them far enough from the parent plant so that the germinating seed won't be shaded by its canopy or have to compete for nutrients in the soil.

Seeds in fluff: Fluff of one sort or another, whether shaped like a parachute or a piece of candy floss, also allows the seed to fly away. Like those with wings, a certain percentage of these seeds will land in a spot where they can sprout and thrive.

Seeds that float: Buoyant seeds come from plants that grow near fresh water or the ocean. Coconut palms on remote islands result from seeds travelling long distances, just as yellow flag (*Iris pseudacorus*) can colonize the edges of a creek. Arrowhead (*Sagittaria* spp.) also has floating seeds.

Seeds as nuts: Nuts can be buried by forgetful squirrels or protected by their hard shells from animals that might otherwise eat them. After a winter or two, the shells split open or degrade, and the nut can begin growth.

Cones protect the seeds from predators.

Seeds in fruit: Fruit attracts insects, birds and animals, all of which are excellent transportation vehicles for the seeds inside. These seeds either require a trip through a digestive system before they can germinate or aren't harmed by it.

Seeds that pop: Popping pods send seeds flying in all directions. The harder they pop, the further the seeds travel. If you look closely at these seeds, you'll see that they're all designed to roll or stick once they've landed.

Seeds in cones: Cones protect the seeds from predators and are heavy enough so that gravity will carry them down any slope where they land, spewing seeds as they go and eventually setting in depressions in the soil.

Seeds that blow: Blowing seeds are so light that they don't need hair or fluff to travel—a passing breeze can carry them away. They lodge in niches in the soil surface, where they'll get adequate light to germinate.

Seeds that stick: Sticking seeds grab on to animals' coats and your clothing for a free ride to a new spot. Along the way, some drop off as they're brushed against other plants, while others are deposited where you or your dog stop to pick them off. An adhesive substance covers mistletoe seeds and "glues" them to the beaks of birds that try to eat them. The birds wipe off the seeds on tree branches, and the parasitic plant has a place to germinate and grow.

Seeds that prick: Prickly seeds discourage predators. If the seed is enclosed in a hull with sharp prickles on the outside, animals aren't likely to penetrate the hull. By the time the hull opens on its own, the seed may be in a good germinating spot.

Seeds for fire: Fire is a natural part of many ecosystems; as a consequence, the seeds of some plants—often called fire followers—require heat or the chemicals left behind in the ash of burned vegetation in order to germinate.

Seeds carried by ants: Ants carry seeds from one site to another, removing them from competition with their parents and broadening ranges.

A Helianthus head is made up of more than 1,000 flowers.

Flowers and seeds

A flower's only purpose, at least from a natural perspective, is to produce seeds. Colours, markings, fragrance, shape – all the things we love and admire about a bloom – are simply attractants. They lure insects and animals to probe into a flower, generally looking for nectar and pollen, and in so doing fertilize it.

Sex is at the bottom of all this, of course. All seeds are formed by a sexual union of reproductive cells that we term as being either female or male. As with human reproduction, a male cell (a sperm cell or male gamete) and a female cell (an egg or female gamete) must unite to form the first cell of the new organism – the zygote. The male gamete is contained in the pollen of the plant; a female gamete is an egg, deep inside the ovary. In the case of plants, when these cells unite, the rapid cell division of the zygote forms the structures that turn it into a seed.

These Begonias are hybrids, so you can't be sure of what their seeds will produce.

Flower forms

As shown in the illustration below, flowers are composed of distinct parts. The flower stalk is called a pedicel. It supports the receptacle, which is actually the top of the stem. Taken together, the receptacle and the sepals – which are green in most plants – are called the calyx. The sepals enclose the flower bud while it's forming, and in most plants open and fall backward to allow the petals to emerge. Tulips are one flower without sepals; as a consequence, their petals are technically known as tepals. The petals or tepals make up the corolla, and the corolla and the calyx together are the perianth.

The male structures of the flower are called stamens and consist of a stalk, or filament, with an anther at its tip. Pollen forms inside the anther and is like the fingerprint of the plant species – no two species have similarly formed pollen grains.

The female reproductive structures include the ovary, the style and the stigma. Taken together, these structures are known as the pistil.

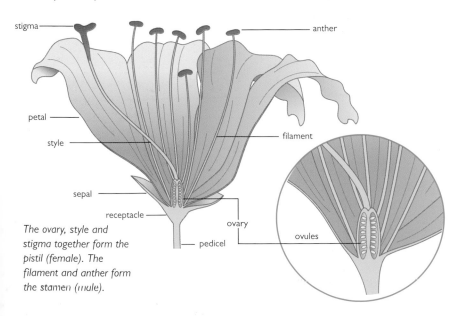

The ovary, style and stigma together form the pistil (female). The filament and anther form the stamen (male).

How it works

When pollen is released from an anther, some of it may land on the stigma, which is sticky. Each grain of pollen contains two cells. One of these cells forms the hollow pollen tube that grows down the style and into a tiny hole in the ovule, and the other divides to become two sperm cells. The sperm cells travel through the pollen tube and into the ovule, where one sperm cell unites with the egg. But the other sperm cell isn't extra – it unites with another cell in the ovule and forms the endosperm, a tissue that stores food to nourish the zygote while it develops into the embryo, the tiny plant that's inside the seed. Some embryos – those in peas, beans and maize seeds, for example – don't use all the endosperm while they are developing into seeds; instead, the remaining endosperm supplies the germinating seed with nourishment. The ovary, which encloses the seeds, becomes the fruit.

This may seem too academic to worry about. However, knowing these simple facts makes it easier to save your own seed or breed cultivars to better suit your environment.

Variations on a theme

When saving your own seed, you'll have to learn whether a plant has perfect or imperfect flowers. A perfect flower is one that contains both male and female parts. The generic flower illustrated on the facing page is perfect: It contains both the pistil and the stamen.

Imperfect flowers, such as begonia blooms, contain only one type of reproductive structure. You can probably name a few vegetables that fall into this category, such as squash and corn, but flowers of many ornamentals do as well.

When both male and female flowers grow on the same plant (as on squash and maize), the plant is called monoecious. But if the individual plants have only one type of flower (as is common on kiwi vines and holly), they are called dioecious. Some plants produce seeds without being fertilized. This is called parthenocarpic reproduction and usually results in a fruit with underdeveloped seeds, such as a "seedless" cucumber.

Above: You can see developing seeds in the bottom calyxes on the stalk. Below: Phlomis flowers are perfect, with male and female parts.

Starting seeds

In the following pages, you'll learn about a number of different treatments that ensure maximum germination rates for particular seeds. Each of these treatments makes a significant difference to your seed-starting success, but it's equally important to know some basic seed-starting techniques.

No matter what other techniques you are using – leaching, stratifying, scarifying, or soaking – the moment will come when you have to put the seed in its starting medium and let it germinate. You'll have a number of choices and decisions to make at each step of this operation, so you may want to review the following before you start a large number of seeds.

Starting inside

Starting seeds inside allows you to grow many plants that might not have time to bloom in your short season, as well as to pamper those that require special treatment while they're germinating or first growing.

Your first task is choosing a starting medium. As a general rule, start seeds for perennials in a compost-based soil mix. If you are concerned about their getting a damping-off fungus, take the steps below. Many perennial seedlings will grow at temperatures below those that favour the fungi and can be started successfully in a cold frame.

Spacing seeds correctly is important in any starting location, inside or out. If you shake tiny seeds straight from the packet, large numbers may fall in the same area. When they are ready to transplant, you'll have a hard time separating their roots without damaging them. So it's worth taking the time to space seeds correctly in the first place.

Three methods are effective. You can buy a small hand-seeder at any garden supply and use it to drop the seeds along the furrow. You can also mix the seeds with about twice as much fine, washed sand or vermiculite and drop tiny pinches of this mixture at

Seed packets are notoriously easy to spill. Remember to set open packets on a saucer so you can pick up spilled seeds.

intervals along the furrow. Finally, you can use a professional grower's technique: spread out the seeds on a plate, moisten the square end of an unfinished wooden chopstick with water, and place it on one of the seeds. It will stick to the chopstick. Place it where you want it in the furrow and press it down – the seed will stick to the soil mix or vermiculite there because the medium is wetter than the chopstick. This system allows you to place seeds with precision.

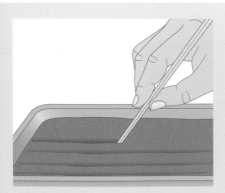
1 FURROW
Make 1cm furrows in the mix.

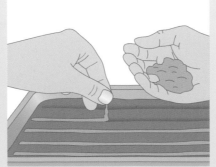
2 FILL
Fill the furrows with fine- or medium-grade gardener's vermiculite.

3 SPACE
Plant the seeds on the top of the furrows.

Cover starting trays or pots with kitchen-grade clingfilm to retain soil moisture. If your seeds germinate within a week, you won't have to water them until after the seeds have germinated and you've removed the covering. But remember not to put a plastic-covered container on a windowsill or under lights – the light can raise the temperature so much that it kills the seeds by baking or steaming!

Location of a seed-starting area is crucial. Start seeds somewhere a little water can spill without causing any damage. If you're using a heating mat, as described on page 18, or supplemental lighting, you'll need an electrical outlet nearby.

Monitor your seeded trays and pots to make certain that the soil temperature is within the preferred range, the moisture is sufficient and the seeds haven't yet germinated. If you work away from home, you can monitor twice a day, but to be safe, start your plants on a weekend so you can monitor soil temperatures in the warmest and brightest parts of the day.

Supplemental lights for young seedlings can be as simple as a fluorescent fixture fitted with standard cool white bulbs. Plants don't need full-spectrum lights until they are beyond the seedling stage. If you are going to light your seedlings – either to extend the hours a day they see light or to increase the light intensity – arrange the fixture so you can raise and lower it or the plants. The tops of the leaves should be about 13cm below the lights at all times.

Move germinated seedling trays to an area where they get adequate light and correct temperatures. Transplant the seedlings into larger growing containers filled with a nutrient-rich mix when their first true leaves – the second set they get – are fully expanded. Note that using unsterilized soil in a seedling mix for indoor growing is asking for trouble. Clean compost is okay to add

When you transplant the seedlings, touch only the root balls and seed leaves, not the true leaves or stems. Otherwise, you could damage the vessels they rely on for water and nutrient transport.

Starting seeds outside

You can start plants either where they are to grow – as you often do with vegetable seeds – or in special nursery beds tucked into unobtrusive spots in the garden. In either area, you'll use some of the techniques discussed above. For example, if you are planting very small seed, it's useful to mix it with fine sand or vermiculite before releasing small pinches along the garden furrow. Similarly, if you are worried that your seed will dry out before it germinates, cover it with vermiculite, water it with a fine misting nozzle, and cover it with horticultural fleece or layers of newspaper, depending on the seed's need for darkness or light. (See pages 34–35 for light/dark information.)

Nursery beds give young plants a fine home in which to get started. It's wise to have a couple of these for different purposes. One can be set aside to grow first-year biennials – in autumn or, in some cases, the following spring, you'll move these plants to their permanent places in the garden. Another can be set aside for young trees and shrubs; once these plants are sufficiently large, you can transplant them into place, but until then they can remain protected in the nursery bed.

Make sure that the soil in these beds has good texture and balanced nutrient levels. You'll want to give your young plants the best possible start in life.

4 MIST
Mist them with a hand-mister.

5 SCATTER
Scatter some more vermiculite on top of the seed tray, and mist again.

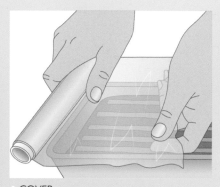

6 COVER
Cover the seed tray or pot with plastic film, and place it in its starting position.

Leaching

Plants have many survival mechanisms, all designed to protect them from harm. Because seeds are so vulnerable, they display some of the most effective of these. Hard shells and barbs are easy to see, but chemicals on the seedcoat aren't apparent to the gardener's eye.

In cold climates, if seeds germinate early in the spring before the weather has settled, the young plants can be susceptible to cold damage or, in the worst cases, freezing. Similarly, if plants in desert environments germinate after there's been enough rain to penetrate their seedcoats but not enough to build up soil moisture, the seedlings could face an arid, unfriendly world.

Fortunately, there is a natural mechanism to protect these plants.

When the seed develops, a chemical layer that inhibits germination forms on top of the seedcoat. This chemical, generally a phenolic compound, is water soluble. However, it is thick enough so that it won't wash off until it has been through many rains.

Do it yourself

You can simulate rain and speed up germination for some seeds. If you practise this technique, you'll cut germination time from 3 weeks to 1 week. Try it with parsley (*Petroselinum crispum*), carrot (*Daucus carota*), and passionflower (*Passiflora incarnata*).

APPROPRIATE PLANTS

In addition to using this technique for the plants listed below, you can try it if you're having trouble with a cool-climate wildflower that's termed an ephemeral or a desert plant that's slow to germinate.

COMMON	BOTANICAL
Mohave Verbena	*Abronia pogonantha*
Desert Sand Verbena	*Abronia villosa*
Desert Paintbrush	*Castilleja chromosa*
Australian Fuchsias	*Correa* spp.
Mohave Aster	*Machaeranthera tortifolia* (syn. *Xylorhiza*)
Waxflowers	*Philotheca* spp. (formerly *Eriostemon*)

Soak seeds overnight.

What can go wrong

Fungal diseases: Wet seeds in a tightly closed container are susceptible to fungal diseases. Remove them from the jar or plastic bag every morning. If you are likely to forget, use an unsealed plastic bag rather than a sealed container to retain moisture.

Death from drought: Seeds will die if they begin to take in moisture and then dry out. If you will be away from the house for more than a couple of hours, pop a plastic bag over the strainer to retain moisture.

Strain and leave the seeds, but do not allow them to dry out.

Checklist

Season: Spring

Tools: Small-mesh strainer

Equipment: Jar or cup for dripping water, jam jar

Supplies: Cotton wool balls, sealing plastic bag, nonsealing plastic bag, water

Temperature: Water temperature should be room temperature, not too warm and not too cool.

1 PREPARE
Place seeds into a jar or other container.

If soaked seeds dry out, they will almost certainly die.

2 SOAK
Cover the seeds with water and let them soak overnight.

3 STRAIN
Drain water through a strainer the next morning. Set the strainer over the jar and leave it close to the sink so you'll remoisten the seeds frequently.

4 LEACH
Run water over the seeds every couple of hours during the day to wash off the inhibiting compounds. Do not let them dry out.

5 COVER
Place the strainer with the seeds in it in a plastic bag along with a couple of moistened cotton wool balls. Seal the bag. This keeps seeds moist enough all night.

6 REPEAT AND REPEAT
Pour water through the seeds the next morning, and repeat this every 2 hours or so. Hold them overnight, as directed above, and plant the following day.

Soaking

Just as some seeds protect themselves with compounds that must be washed off the seedcoat before they'll germinate, others do so by being resistant to taking in water. Without water, the seed can't germinate. But it's easy to get around this problem – simply soak the seeds for anywhere from a few hours to a week.

Seeds that require soaking usually have thick, tough seedcoats. You might imagine that the plants that produce these seeds live in swamps or thrive in consistently damp soil. But this isn't always so. For example, thrift (*Armeria* spp.) and spurge (*Euphorbia* spp.) are excellent choices for a dry, rocky soil. Again, this adaptation simply protects seeds from germinating before the soil is adequately moist to sustain their early, fast growth.

Many seeds that need soaking also require another treatment, such as stratifying (see page 32) or scarifying (see page 30). Always check to see if the seed you are germinating is listed in any other categories. If your seeds require both of the above treatments, do them in this order: chill or freeze the seeds first, nick or scratch them next and then soak them. The seeds may need light or dark to germinate, so check those lists as well.

Canna seeds benefit from soaking.

Checklist

Season: Spring

Tools: Small-mesh strainer

Equipment: Ramekin or small bowl, paper coffee filter

Supplies: Water

Temperature: Unless otherwise indicated, water temperature should be room temperature, not too warm and not too cool.

Unbleached coffee filters make excellent seed strainers – not even the tiniest seed can slip through.

What can go wrong

No germination: Timing is everything when it comes to soaking seeds. If you look at the list at right, you'll see very specific directions. Someone – maybe a researcher, maybe a gardener – took the time to figure out just how long a seed should soak. Follow these guidelines for best results.

Fungal diseases: If the seed soaks too long, it will rot. Again, pay attention to the directions at right. If you are trying to germinate a seed that's not on the list and don't have specific directions, err on the side of caution and soak it overnight, 8 hours maximum.

1 SELECT SEEDS
Place the seeds you want to soak in a ramekin or small bowl.

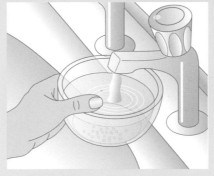

2 ADD WATER
Let the seeds soak for the appropriate amount of time.

3 STRAIN
At least once every 12 hours, pour off water. Use a coffee filter to capture seeds.

4 SOAK OR PLANT
Follow directions to keep soaking or plant immediately.

You'll notice that seeds swell as they are soaking.

If you soak seeds too long, you could end up with a mould problem such as this.

APPROPRIATE PLANTS

COMMON	BOTANICAL	TREATMENT
Musk Mallow	*Abelmoschus moschatus*	After scarifying, soak 1 hour
Thrift	*Armeria* spp.	Soak 6 hours, warm water
Canna	*Canna* x *generalis*	After scarifying, soak 48 hours, change water every 12 hours
Spurge	*Euphorbia* spp.	After stratifying, soak 5 days
Globe Amaranth	*Gomphrena globosa*	Soak 3 days, warm water
Daylily	*Hemerocallis* spp.	Freeze 2 weeks, soak 5 days
Hibiscus	*Hibiscus* spp.	Soak 48 hours, change water every 12 hours
Sweet Pea	*Lathyrus odoratus*	After stratifying, soak 48 hours, change water every 12 hours
Lupin	*Lupinus polyphyllus*	Soak 2–3 days
Bells of Ireland	*Moluccella laevis*	Soak 24 hours

Scarification

Scarification happens naturally when a bird swallows a seed that then passes through the digestive system. Acids there eat through the tough seedcoat. When the bird eliminates the seed, it's ready to germinate.

APPROPRIATE PLANTS

COMMON	BOTANICAL
Mallow	*Abelmoschus* spp.
Hollyhock	*Alcea rosea*
False Indigo	*Baptisia* spp.
Angel's Trumpet	*Brugmansia* spp.
Canna	*Canna* x *generalis*
Moonflower	*Ipomoea alba*
Morning Glory	*Ipomoea* spp.

It's easy to remember what scarification means because the root of the word is "scar", and that's what you do when you scarify a seed – you scar the seedcoat. Many seeds that birds inadvertently plant can either tolerate scarification or require it, but this is not always true. For example, birds are responsible for seeding many thousands of multiflora roses *(Rosa multiflora)* every year, but the roses you start from seed don't need scarification. Scarifying a seed that doesn't require it can kill it, so it's best to check a good reference if you have a tough seed that looks as if it might need this treatment but isn't listed on this page.

This procedure might alarm you the first time you do it, but it's really not frightening. You simply want to scratch the seedcoat just enough so that water can penetrate it. As shown right, you can do this by nicking the seedcoat with the tip of a very sharp, small paring knife or by scratching it with sandpaper. Nick or scratch the side of the seed to avoid injuring the hilum, or "eye" of the seed, where the first root and shoot will emerge.

Scarified and soaked Ipomoea seeds tend to grow well.

Birds love Abelmoschus seeds and can inadvertently plant them in their travels.

Fine sandpaper is best for scratching seeds.

1 SELECT SEEDS
Pour seeds into a bowl to examine them.

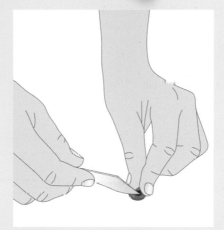

2 NICK
Use a very sharp knife to nick the seedcoat. Do not injure the hilum, or "eye", because the first root and stem emerge from this area.

3 SORT
Place the nicked seed in another bowl to keep your piles separate. Scarify the remaining seeds, one by one.

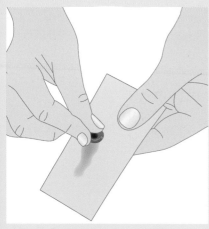

ALTERNATIVE METHOD: SANDPAPER
Alternatively, scarify seeds by rubbing them on a piece of fine sandpaper. If you are working with large seeds, simply draw each seed over the sandpaper, but if you are working with a group of seeds, rub them between two pieces.

Checklist

Season: Spring

Tools: Small, sharp paring knife

Equipment: 2 small bowls

Supplies: Seeds, fine sandpaper

Sandpaper is a good choice with Alcea rosea seeds, above right, and also does a nice job on Lathyrus odoratus, right.

Stratification

In cold climates, seeds germinate in spring. This gives them the whole growing season to prepare for winter, either by setting and ripening seed before they die if they are annuals or, if they are perennials, by dying back or hardening their tissues before the cold sets in. Seeds of these plants have developed a way to protect themselves from germinating too early or too late – they must go through a certain period of chilling before they can germinate.

When a gardener purposely chills seeds, we call it stratification. It was once believed that certain seeds needed freezing before they would germinate, but recent research indicates that chilling will do. Try both to discover which works best. Put the seeds listed in the chart below as needing freezing in the refrigerator for an equivalent amount of time.

Stratifying seeds is one of the easiest gardening techniques you'll ever do. In most cases, after you buy your seed packets in late winter, simply pop the unopened packets of the ones you'll stratify into a jar with a lid that closes tightly. Place the jar in the freezer or the refrigerator. To avoid confusion, it's best to use a waterproof marker to label each packet with the earliest date that you could remove it – it doesn't hurt to let it stratify for longer than the recommended time, but it does to reduce it.

Some seeds – particularly those of cold-loving conifers – have the highest germination rates if they experience alternate freezing and thawing. If you live in a climate that is reliably cold all winter, the easiest way to stratify these seeds is to plant them in seed trays in late autumn and place them in a cold frame with an automatic vent opener. Steady cold followed by spring temperature fluctuations will stimulate their germination, just as it does in the wild.

APPROPRIATE PLANTS

COMMON	BOTANICAL	TREATMENT
Monkshoods	Aconitum spp.	Chill 3 weeks
Snapdragon	Antirrhinum majus	Chill 4–6 weeks
Columbines	Aquilegia spp.	Chill 2–8 weeks
Blackberry Lily	Belamcanda chinensis	Chill 4–6 weeks
Clematis	Clematis spp.	Chill 3 months
Larkspurs	Consolida spp.	Chill 6 weeks
Delphinium	Delphinium elatum	Freeze 4–6 weeks
Pinks	Dianthus spp.	Chill 4–8 weeks
Bleeding Heart	Dicentra spectabilis	Chill 6 weeks
Dragon's Heads	Dracocephalum spp.	Chill 4–6 weeks
Coneflowers	Echinacea spp.	Chill 3–6 weeks
Sea Hollies	Eryngium spp.	Chill 6 weeks
Spurge	Euphorbia griffithii	Freeze 2 weeks, soak 5 days
Gentians	Gentiana spp.	Chill 2–4 weeks
Cranesbill	Geranium sanguineum	Chill 1–2 months
Oxeye	Heliopsis helianthoides	Chill 4 weeks
Christmas Rose	Helleborus niger	Freeze 2 weeks
Daylilies	Hemerocallis spp.	Freeze 2 weeks, soak 5 days
Sweet Pea	Lathyrus odoratus	Chill 2–3 weeks, soak 48 hours
Lavenders	Lavandula spp.	Chill 4 weeks
Gayfeather	Liatris scariosa	Chill 6 weeks
Cardinal Flower	Lobelia cardinalis	Chill 10 days
Monkey Flowers	Mimulus spp.	Chill 3 weeks
Bells of Ireland	Moluccella laevis	Freeze 5 days
Missouri Primrose	Oenothera macrocarpa	Chill 2 weeks
Peonies	Paeonia spp. & cvs.	Chill 2 months
Penstemons	Penstemon spp.	Chill 4–8 weeks
Phloxs	Phlox spp.	Freeze 2 weeks, soak 7 days, change water daily
Jacob's Ladder	Polemonium boreale	Chill 2 months
Primrose	Primula marginata	Chill 2 weeks
Roses	Rosa spp. & cvs.	Chill 5 months
Black-Eyed Susan	Rudbeckia hirta	Freeze 1 week, chill 1 week
Salvias	Salvia spp.	Chill 1 week
Canadian Burnet	Sanguisorba canadensis	Chill 6 weeks
Globeflower	Trollius europaeus	Freeze 2 weeks
Veronicas	Veronica spp.	Chill 2 months
Pansies, Violets	Viola spp.	Chill 4 weeks
Adam's Needle	Yucca filamentosa	Freeze 3–4 weeks

What can go wrong

Poor germination: Several factors can lead to poor germination of stratified seeds. You may not have given them a long enough period of chilling. If you suspect this is the case, wrap the entire seed tray in plastic and pop it in the refrigerator for another period of stratification.

Drying out: The seeds could also have dried out. Because all but the very oldest refrigerators pull humidity from the inside air, it's important to put the seed packets in a sealed plastic bag or a closed glass jar.

Label your packet.

Checklist

Season: Spring

Equipment: Jar with tightly fitting lid, seed trays

Supplies: Seeds, fine-point waterproof pen, potting soil, fine sandpaper

Temperature: Chill at 4°C, the temperature that most domestic refrigerators maintain. If you own a chest or upright freezer, use it for freezing seeds. If not, place seeds in the centre of the freezer compartment of your refrigerator.

1 LABEL
Label the seed packet with the date to remove it from the cold.

2 SEAL
Put the packet into a glass jar or plastic bag you can seal. Seal and place in the refrigerator.

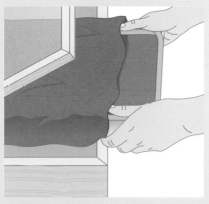

ALTERNATE METHOD: 1 COVER
Many seeds that require chilling do best in a tray in a cold frame over winter, and some require darkness for germination. If so, cover the seeded tray with black plastic before placing it in a cold frame.

2 LAYER
A thick layer of newspaper over the black plastic further excludes light, and the fluctuating temperatures in the cold frame will stimulate germination.

Enclose seeded trays or cell trays in a plastic bag you can seal when you put them in the refrigerator or freezer. If the bag is not sealed, moisture will evaporate and the seeds will die of drought.

Light

Tiny seeds have tiny shoots and first roots; if they germinate deep in the soil, the shoots can't grow long enough to get the seed leaves above soil level. Without exposure to light, these leaves can't manufacture the sugars that will allow the tiny plant to keep growing. Nature's solution to this dilemma is to make certain the seed can't germinate unless it's close to the soil surface.

Whenever you are starting small seeds, you can expect that they will need light to germinate. But, as you'll see from the list at right, a few larger seeds such as hollyhock (*Alcea rosea*) and coneflower (*Echinacea* spp.) must also see light to germinate well.

Place a sheet of polythene on top of the seed tray and set under a fluorescent fixture.

What can go wrong

Poor germination: If seeds slip too far into a niche in the soil mix, they may not see enough light to germinate. To prevent this, gently pat and smooth the mix before you seed. Seeds on the surface can also dry out; make certain to cover the medium and keep it moist while seeds are germinating.

Fungal diseases: Check the trays several times a day, and remove the plastic covering as soon as you see tiny white roots or green leaves—otherwise the lack of air circulation around the seedling could lead to rot.

APPROPRIATE PLANTS

COMMON	BOTANICAL
Ageratums	*Ageratum* spp.
Hollyhock	*Alcea rosea*
Bishop's Flower	*Ammi majus*
Snapdragon	*Antirrhinum majus*
Columbines	*Aquilegia* spp.
Thrifts	*Armeria* spp.
Wormwoods	*Artemisia* spp.
Butterfly Weed	*Asclepias tuberosa*
Wax Begonias	*Begonia* Semperflorens group
Strawflower	*Bracteantha bracteatum*
Celosias	*Celosia* spp.
Pinks, Sweet Williams	*Dianthus* spp.
Foxglove	*Digitalis purpurea*
Coneflowers	*Echinacea* spp.
Globe Thistle	*Echinops ritro*
Eucalyptus	*Eucalyptus gunnii*
Blanket Flower	*Gaillardia aristata*
Baby's Breaths	*Gypsophila* spp.
Sneezeweed	*Helenium autumnale*
Dames Rocket	*Hesperis matronalis*
Coral Bells	*Heuchera* cvs.
Annual Candytuft	*Iberis amara*
Impatiens	*Impatiens walleriana*
Annual Statice	*Limonium sinuatum*
Lobelias	*Lobelia* spp.
Honesty	*Lunaria annua*
Forget-Me-Not	*Myosotis* spp.
Missouri Primrose	*Oenothera macrocarpa*
Petunia	*Petunia hybrida*
Chinese Lantern	*Physalis alkekengi*
Balloon Flower	*Platycodon grandiflorus*
Primroses	*Primula* spp.
Salvias	*Salvia* spp.
Pincushion Flowers	*Scabiosa* spp.
Feverfew	*Tanacetum parthenium*

Checklist

Season: Spring

Tools: Unfinished wooden chopstick or small hand-seeder

Equipment: Seed tray or cell tray, wide windowsill or fluorescent plant light fixture

Supplies: Seeds, seed-starting medium, clear plastic wrap

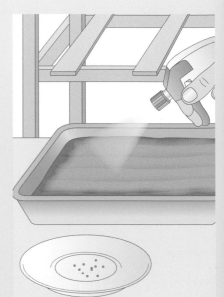

Place seeds that require light on the surface of the medium, and mist them into niches where they'll stay moist and still "see" light.

Dark

Some seeds simply won't germinate while they are exposed to light. This survival mechanism safeguards the seed from germinating too close to the soil surface, where moisture supplies can be too erratic to sustain the seed while the first root is elongating.

Lupin seeds require moist, dark conditions for germination.

The mechanism that controls this response is a light-sensitive chemical on the seedcoat. It registers light intensity and inhibits germination when it is too great. But seeds that germinate too deep in the soil will die, too, because their shoots won't be able to push the seed leaves into the light. For that reason, these seeds often require a certain ratio of oxygen to carbon dioxide; deep under the ground, the ratio of CO_2 to oxygen is much higher than it is near the soil surface. Understanding this helps when you start these plants. Outside, cover them just enough to exclude light; inside, cover the flats with a layer of newspaper rather than wrapping them tightly in plastic that excludes air exchange.

APPROPRIATE PLANTS

COMMON	BOTANICAL
Pot Marigold	Calendula officinalis
Cornflower,	
Bachelor's Button	Centaurea cyanus
Larkspur	Consolida ambigua
Delphinium	Delphinium elatum
Roman Shields	Fibigia clypeata
African Daisy	Gazania linearis
German Statice	Goniolimon tataricum
Sweet Pea	Lathyrus odoratus
Lupins	Lupinus spp.
Phlox	Phlox spp.
Moss Roses	Portulaca spp.
Painted Tongue	Salpiglossis sinuata
Butterfly Flower	Schizanthus x wisetonensis
Verbenas	Verbena spp. & hybrids
Pansy	Viola x wittrockiana

What can go wrong

Poor germination: If seeds aren't covered well enough, they may not germinate. Bury the seed to a depth about three times its width—no more—and cover the flat with newspaper. Outdoors, use an old board or weighted cardboard to cover the seed furrows.

Fungal diseases: Check the seeds several times a day, and remove the covering as soon as you see the tiny seedlings; otherwise the lack of air circulation around them could lead to rot.

A sheet of black plastic will keep out light, but use newspaper over the seeds to avoid rotting.

Checklist

Season: Spring

Tools: Unfinished wooden chopstick or small hand-seeder

Equipment: Seed tray or cell tray

Supplies: Seeds, seed-starting medium, newspaper

Place a half-inch layer of newspaper on top of the seeded tray. The newspaper lets the soil mix "breathe" while also excluding light.

Relative humidity

Relative humidity levels don't affect seeds in a moist soil mix, but they do affect seedlings as soon as they germinate and push their first shoots and seed leaves into the surrounding air. Relative humidity levels affect cuttings, too, so it's useful to learn how to manipulate them.

Relative humidity is a measure of the amount of moisture in the air, compared with the amount of moisture that the air could hold at that temperature. Molecules in warm air move more swiftly than molecules in cold air, so warm air can hold more moisture than cold air – the faster movement of particles means that there's more room for water vapour.

Plants transpire, or give off water vapour, much in the way that people perspire. The surrounding air takes up the released water vapour, and the roots take up moisture from the soil to replace the lost moisture. If the air contains a high relative humidity, it cannot take in much from the plant, so the leaves are bathed in a thin layer of water vapour. Many disease-causing fungal spores germinate in these conditions, so it's hard to keep plants healthy if the relative humidity is too high. But if the relative humidity is too low – and particularly if the air is warm – plants suffer from lack of adequate moisture. Cuttings are particularly vulnerable to low relative humidity conditions because they have few, if any, roots to resupply water.

The high humidity in a rain forest determines which plants can thrive.

APPROPRIATE PLANTS FOR HIGH RELATIVE HUMIDITY

Although the majority of plants thrive in relative humidity levels of about 50 to 60 per cent, plants native to tropical and subtropical regions require levels between 65 and 85 per cent at temperatures ranging between 24° and 32°C. Many houseplants, whether foliage or flowering, fall into this group, and some, such as warm-temperature orchids, suffer below 70 per cent.

COMMON	BOTANICAL
Begonias	*Begonia* spp.
Peacock Plants	*Calathea* spp.
Orchids, including	*Cattleya, Dendrobium, Oncidium, Phalaenopsis,* and *Vanda* spp.
Croton	*Codiaeum*
Nerve Plants	*Fittonia* spp.
Fuchsias	*Fuchsia* spp.
Azaleas	*Rhododendron* spp.
Bromeliads	*Vriesea, Billbergia, Aechmea, Ananas, Bromelia, Guzmania, Cryptanthus* and *Neoregelia* spp

APPROPRIATE PLANTS FOR LOW RELATIVE HUMIDITY

Plants that prefer low relative humidity levels – about 35 to 45 per cent – generally hail from desert areas or sunny, south-facing rocky slopes. They suffer when relative humidity levels are 20 per cent or less. This may sound very low, but remember that without a humidifier, even homes in high-rainfall areas experience relative humidity levels of 15 to 20 per cent when they are heated during the winter months.

COMMON	BOTANICAL
Agaves	*Agave* spp.
Aloe	*Aloe* spp.
Rock Purslane	*Calandrinia* spp.
Cacti, including	*Cephalocereus, Cereus, Echinocereus, Ferocactus, Lemaireocereus, Opuntia, Pereskia, Rhipsalis, Selenicereus,* and *Trichocereus* spp.
Kangaroo Vine	*Cissus antarctica*
Eucalypti	*Eucalyptus* spp.
Spurges	*Euphorbia* spp.
Silk Oaks, Spider Flowers	*Grevillea* spp.
Tea Trees	*Leptospermum* spp.
Rosemary	*Rosmarinus officinalis*
Hens and Chicks	*Sempervivum* spp.
Yuccas	*Yucca* spp.

Checklist

Season: Anytime you are growing seedlings or propagating with cuttings

Equipment: Propagation chamber, pebble trays, fans, humidifier

Supplies: Plastic sheeting, narrow planks to support plastic, long-fibred peat moss

Pebbles provide useful drainage beneath flats and increase the level of humidity.

What can go wrong

Humidity too low: Warning signs include dry, browning leaf tips and margins; dropped lower leaves and flowers that fade too quickly; poor growth; lacklustre leaf colour and sheen; and general ill health. Drape plastic sheeting over groups of plants to increase relative humidity, or grow them in a propagating chamber.

Humidity too high: Fungal diseases – including powdery and downy mildew, botrytis, and those characterized by spots and blotches on the leaves – are an indication that humidity levels are excessive. Other signs include excessively large, floppy leaves and long internodes. Remove coverings as soon as seedlings germinate, and ventilate tented seedlings and propagation chambers to keep humidity levels at 75 to 80 per cent.

RETAIN HUMIDITY

This homemade germination enclosure, shown before the addition of plastic over the sides and rear, retains humidity when the plastic door is down but releases it when the door is raised.

EVAPORATION HELPS

Water evaporating from a pebble tray increases the humidity around seedlings. Monitor to make certain that humidity levels are not too high.

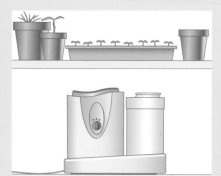

ELECTRIC AID

Use an electric humidifier if your heated home has relative humidity levels below 30 per cent.

TRY AUTOMATION

A commercially available propagation chamber creates an ideal environment but is too expensive for anyone who doesn't start hundreds of cuttings a year.

SLOW AND STEADY

Moistened peat moss in the outer pot slowly releases a steady amount of humidity into the air, keeping cuttings moist.

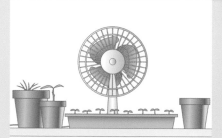

BLOW IT AWAY

When relative humidity levels are too high, break up the boundary layer around plant leaves by using a fan to blow it, and the humidity it contains, away.

Temperatures

Temperature is one of the most important elements in techniques as varied as rooting cuttings, air layering and grafting. But it's crucial when it comes to germinating seeds – temperature can make the difference between success and failure.

Consistent air temperature is important to a seedling once it has germinated.

Make sure plastic sheeting is lifted a bit above the soil.

Two temperatures are important: air and soil. The soil temperature affects the rate of germination – more seeds of a particular species germinate when the soil is within their preferred temperature range than do when it is too cold or even too warm. In general, you get the best germination when the soil temperature is about 5 degrees higher than the plant's ideal air temperature. For example, even though columbines (*Aquilegia* spp.) grow well in cool spring temperatures of 10° to 15°C, their seeds germinate best at 18° to 24°C. Certain warm-weather plants such as coleus (*Solenostemon*) germinate well at their ideal air temperature (27° to 29°C). Air temperature is important once a seed has germinated and when you are rooting cuttings or making grafts.

Increase soil temperatures with a heating mat. If you want to increase air temperatures, enclose the heating mat and flats in a structure made from lathing strips and plastic sheeting. Ventilate as necessary to keep both relative humidity levels and air temperatures within an appropriate range.

Small meat thermometers make good soil thermometers for trays.

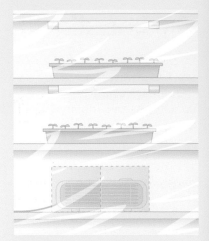

IDEAL SPACE
Create an ideal growing space by supplying light and heat and using plastic sheeting to retain humidity. Shield the plastic from the heater with two layers of plywood.

ADD HEAT
Use a heating mat to increase soil temperatures while seeds are germinating.

What can go wrong
Seeds don't germinate: Soil temperatures higher or lower than preferred ranges can kill seeds, and excessively high or low air temperatures can kill seedlings. Use your thermometers to monitor.

Carbon dioxide (CO$_2$) levels

Without carbon dioxide, there would be no life as we know it. Plants use CO$_2$, water and sunlight in cells containing phosphorus to create the simple sugar from which they make all the other compounds they contain. As they do so, they release the oxygen that animals require, along with water.

APPROPRIATE PLANTS
All plants require CO$_2$ levels of at least 300 ppm, and most respond well to levels higher than this. When you are propagating, increase levels in all plastic tents and germination chambers holding plants or stems with leaves. You'll notice that both your cuttings and seedlings grow stronger and more rapidly as a result.

The amount of CO$_2$ in the air is roughly 340 parts per million (ppm). However, as plants in an enclosed space (such as a greenhouse or grow room) use it, levels can decline so much that plants can't get enough to keep making sugars. Their growth slows as a result. It makes sense that they'll be healthier if you keep the CO$_2$ levels at about 300 ppm.

However, you might be surprised to learn that almost all plants grow better when CO$_2$ levels are higher than normal. Greenhouse growers routinely increase the CO$_2$ levels to 800 to 1,000 ppm when the sun is bright, to boost plant growth and yields. To do this, they inject CO$_2$ into the greenhouse atmosphere from special canisters. This is impractical in a small greenhouse or in the home, of course, but there is a simple way to maintain CO$_2$ levels on a small scale.

Plants in a portable greenhouse will benefit if you increase CO$_2$ levels. Add baker's yeast (see below) and zip the door closed to maintain higher levels. Monitor overall heat and humidity.

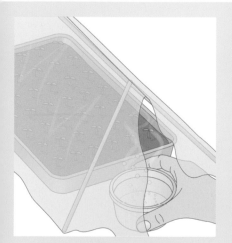

ADD YEAST
Baking yeast gives off carbon dioxide as it grows, so it's one of the easiest ways to increase CO$_2$ levels in an enclosed space. Fill a small bowl half to three-quarters full of warm water, and add a healthy pinch of sugar. Stir it and immediately add a tablespoon, or roughly a packet, of baker's yeast. Place it in the enclosed space, and it will release CO$_2$ as it grows. Test for CO$_2$ after the yeast has stopped bubbling, and repeat as necessary. Please note that in an enclosed space like the one illustrated, the plastic must be closed tightly.

This inexpensive test kit measures CO$_2$ levels from 300 to 5,000 ppm, and you can find it on the Internet. Develop the habit of checking CO$_2$ levels around your plants on sunny mornings.

Life cycles and growth characteristics

Understanding a plant's life cycle and growth characteristics provides vital information when you are propagating it. If you know that it's an annual, for example, you can be certain that the easiest way to propagate it is to save its seeds. But if it's a woody perennial, you might be better off dividing it (if it has a fibrous root system) or taking cuttings. So it's worthwhile reviewing the basics of plant physiology and life cycles.

Annuals

Almost all gardeners grow at least a few annuals, plants that complete an entire life cycle—from seed to seed—in one season. While you can take greenwood cuttings of some annuals, root them, and continue growing them through the rest of the season, this is usually more trouble than it's worth. Starting annual plants from seed is the easiest and most practical way to increase your stock. (You'll learn about simple seed-saving techniques on pages 42–43.)

Winter annuals are easily confused with biennials because their seeds germinate in midsummer; they grow a rosette of leaves before autumn; and then, the following spring, they grow more leaves and a flowerstalk. They are still classified as annuals; however, because they live for 1 year only. It's likely that the only winter annuals you'll ever come across in your garden are weeds—winter cress (*Barbarea verna*) or shepherd's purse (*Capsella bursa-pastoris*).

Tender annuals are intolerant of cool temperatures, and many of them—such as impatiens (*Impatiens walleriana*) and Madagascar periwinkle (*Catharanthus roseus*)—are actually perennials in their native habitats.

Hardy annuals can withstand cool temperatures. Marigold (*Calendula officinalis*) and larkspur (*Consolida ajacis*) are common hardy annuals.

Calendula is a hardy annual that grows well from seed each year.

Biennials

Biennials live for 2 years. Their seeds sprout in the spring or late summer of the first year. They generally grow a crown or rosette of leaves that year, and then, the following season, grow more leaves and, finally, a seed stalk. You will grow them from seed, too, but it's common to grow them in a nursery bed at the back of the garden the first year and then transplant them into their blooming position in the spring of their second year. Canterbury bells (*Campanula medium*) and hollyhocks (*Alcea rosea*) are common biennials.

Perennials

Perennial plants live for 3 or more years. As you can imagine, this is an enormous group of plants that encompasses everything from tiny woodland wildflowers to huge, towering trees. Most perennials make viable seeds, but many do not. And even if a plant does make seeds, it may be easier to propagate it vegetatively—by growing a new individual from a section of the parent plant—than to start it from seed.

Perennials fall into several groups, according to their physical characteristics, and each of these groups is further subdivided, again according to these qualities.

Tender perennials, such as flowering maple (*Abutilon* x *hybrida*) and scented geraniums (*Pelargonium spp.*), are hardy in their native environments but do not survive cold weather and frosts. Gardeners in northern areas typically grow them as houseplants during cool-weather months and allow them to summer outdoors. Propagate according to other growth characteristics, not according to their hardiness.

Hardy perennials, such as coneflower (*Echinacea purpurea*) and roses (*Rosa* spp.), do withstand freezing winters, although they vary in the duration and amount of cold they can survive. Before buying a hardy perennial, always check the "hardiness zone rating" to make sure it will survive in your climate. Again, the propagation technique depends on other growth characteristics of the plant, rather than on hardiness.

Herbaceous perennials have cold-tender stems and leaves that die down at the end of every season in cold climates. Their roots remain alive, although dormant, in the cold or frozen soil, and the plant grows new top growth from the root tissues every spring. A huge number of garden ornamentals fits this description—think of how many of your favourite plants, from anemones (*Anemone* spp.) to yarrows (*Achillea* spp.), have leaves that die down at the end of the season and reappear again the following year. In general, you divide these plants in very early spring, just before they begin growing vigorously, and take cuttings later in the season.

Woody perennials develop woody trunks, stems and branches. Most of the plants you use to form the architecture of your garden design—shrubs, trees and climbing vines—fall into this category. Again, you'll propagate them according to other characteristics—some grow best from seeds, others from suckers or layered branches, and some from cuttings; some must be grafted on to particular types of rootstock.

Broadleaf evergreens are those that keep green leaves throughout the entire year, such as hollies (*Ilex* spp.) and boxwoods (*Buxus* spp.). Propagation techniques for these plants generally include layering, cuttings and occasionally grafts.

Conifers are needle-leafed evergreen plants such as pines (*Pinus* spp.) and spruces (*Picea* spp.). While you can propagate these plants with various other methods, many are easily started from seed.

Natural variations in the colours, shapes and sizes of conifers allow you to create a stunning picture composed only of these plants.

This 'Horsford' white pine adds drama to any landscape.

Saving seeds

Saving seeds opens up one of the most exciting aspects of gardening. Simple techniques allow you to breed plants especially adapted to your garden's environment, learn to create your own hybrids, and breed back from commercially produced hybrids to create open-pollinated plants.

Checklist

Season: Spring to autumn

Tools: Small watercolour brush

Equipment: Wire plant cage

Supplies: Horticultural fleece or geotextile, strong packing tape, string

Saving seeds is thrilling – once you begin, you'll wonder why more people don't do it. The basic techniques of improving a variety are simple and prepare you for the more complicated, but still easily achievable, procedures associated with developing hybrids and breeding back from hybrid cultivars.

Aside from some basic equipment, three things are essential for this work:
• A basic understanding of the biology of seed production
• The patience to continue working on your improvements over the years
• Organizational skills, so that you can keep your seeds sorted and labelled and also maintain accurate records.

Most people begin seed-saving and breeding experiments with vegetable plants, and it's easy to find good books and resources that cover procedures for these. But if you want to work with ornamental plants, don't be dismayed – you can easily adapt the information about vegetable plants to the ornamentals you want to improve.

Distinctions

Teach yourself to save seeds and breed plants by working with an open-pollinated (OP) variety or cultivar first.

Open-pollinated plants come "true to type", which means that plants from their seeds will be very similar to their parents, as long as they are fertilized by pollen from a plant of the same variety or cultivar. In contrast, a hybrid is produced by crossing a species, variety, or cultivar of one plant with another related plant.

It's also useful to know if a plant is a variety or a cultivar. This distinction can be hard to draw because a cultivar is simply a variety that has been selectively bred for its particular characteristics.

Ensure seeds are neatly organized and labelled.

Once these qualities are stable from one generation to the next, the variety has become a cultivar – or cultivated variety. You will be creating cultivars for your own garden.

Isolation

In some cases, all you need to do is collect the dry seedheads or pods at the end of the season and keep them over the winter. But this only works if you and every other gardener in the neighbourhood are growing only one species of an open-pollinated plant – and that's hard to count on.

Some flowers are pollinated by insects or birds that fly many miles in their search for nectar; others are pollinated when wind carries the pollen from one flower to another; still others are self-pollinated when pollen from their anthers drops onto their stigmas. The ease with which pollen travels between plants means that you have to guard against accidental pollination.

The only way to be absolutely sure that you are growing out the seeds you want is to "isolate" the plants. Commercial seed growers do this by putting long distances between different crops. But this is impractical on a home basis, so you'll have to be a little more innovative. As shown in the illustrations at right, there are a number of ways to isolate crops – choose the method that serves you best.

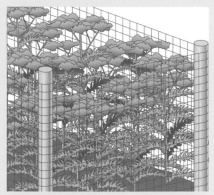

CAGING
Place a cage around the plants you'll isolate long before their blooms open.

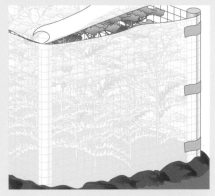

2 ENCLOSING
Use horticultural fleece or geotextile to completely enclose the caged plants. Pile soil on the bottom and put a top on the cage to completely exclude insects or flying pollen.

3 POLLINATING
When blooms open, use a fine brush to move pollen from one flower to another to fertilize the seeds.

ISOLATING WIND POLLINATORS

Many plants, including ornamental grasses, have pollen that floats on the air between one flower and another. Isolate them in cages or make "bags" for their flower heads.

BAG
Surround the flower heads with horticultural fleece, and tie it at the base. Wrap cotton wool around the stems, just under your tie, to exclude stray pollen grains and prevent developing seeds from dropping to the ground.

2 HARVEST
Once the stems are dry and brown, the seeds are ripe. Tip the bag to catch falling seeds and cut the stems. Let the seeds finish drying inside the house once you've collected them.

What can go wrong

Improper pollination: Sometimes you will end up with seeds you never meant to get. This occurs often in the squash family when one of the plants yields fruit with characteristics from two different parents — not the one you wanted.

Too few seeds: Some isolated plants will develop too few seeds unless you hand-pollinate them — remember to carry through with that step.

A few supplies form the basis of this extremely satisfying technique.

Pollination

Taking over nature's role in pollination means that you have to adhere as closely as possible to natural systems, and, when that is impossible, substitute a method that is almost as effective.

Pollination seems very simple: you move pollen grains from the anthers of a flower and place them on the stigma of that flower. For many plants, pollination is just this straightforward – but this isn't always the case.

Some perfect flowers that you might imagine to be self-pollinating are actually self-incompatible. They cannot pollinate themselves because of chemical inhibitors or because they are prevented by timing – their pollen is ripe when their stigmas are not able to accept it. This quality is common in insect- and animal-pollinated flowers, so it's best to assume it's so unless you know differently. This means that when you are saving seeds from most plants, you need to transfer the pollen from flowers on one plant to flowers on a different plant – just as insects do.

Wind-pollinated plants can also be tricky to pollinate. In spinach, for example, plants are either male or female. If you want to save seed, you'll have to cage a large group of plants and remember to shake them every day while the plants are in bloom. Pollen moves most easily when the air is not too humid, and stigmas of day-flowering plants tend to be most receptive between about 10 A.M. and 1 P.M., so it's important to pollinate during these hours.

POLLINATOR CLUES

It's useful to know how a plant is naturally pollinated, and often you can work this out by observation or research. But if you're stumped, use the following guidelines to make a good guess.

POLLINATOR	FLOWER CHARACTERISTICS
Bat	Dull white or green; brushlike or bowl-shaped; musky, fermenting odour; night blooming; nectar producer
Bee	Yellow, blue, or purple; flower parts not recessed; sweet or fresh scent; nectar and/or pollen producer
Beetle	Brown, purple, or dull coloured; flowers flat to bowl-shaped; strong, fruity fragrance; edible flower petals
Bird	Bright red; tubular corolla; usually not fragrant; nectar producer
Butterfly	Red, yellow, blue, or purple; flower parts not recessed; sweet or fresh scent; nectar producer
Fly	Purple or brown; flowers flat or shallow; odour similar to decaying flesh
Moth	White or pale green; narrow corolla; night flowering; strong, sweet scent; nectar producer
Wind	Pale green or yellow, small, inconspicuous flowers; often growing on flexible stems; very light fragrance

Flowers on plants such as sweet peas (*Lathryus odoratus*) and tomatoes self-pollinate before they are fully open. But accidental insect-pollination is possible. To protect against random pollination, you can "bag" individual blooms, as described on page 43. But rather than waiting until they are pollinated, bag them a few days before they open. To assure fertilization, shake the stems every morning from the time you bag them until the petals drop.

Unlike many other flowers, sweet peas will self-pollinate.

1 SELECT PLANTS
Choose the plants you want to breed. In this case, the gardener wants to develop calliopsis (*Coreopsis tinctoria*) flowers with a large central coloured area rather than a small one.

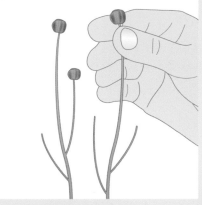

2 FIRST GENERATION
Begin by using coloured electrical tape to mark the stems of plants with the desired characteristics. Save seed from them to grow out the following year.

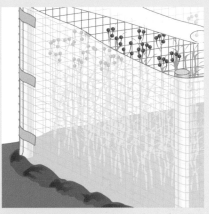

3 SECOND GENERATION
The next year, cage and enclose the plants you grow from your saved seed.

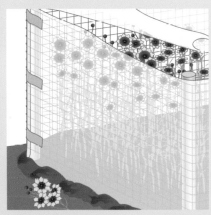

4 SELECT PLANTS
Once plants bloom, pull up those without the desired characteristics but keep those with the desirables. Hand-pollinate the flowers.

5 THIRD GENERATION
By the third year, many more plants will have the desired characteristics. Pull out those without and pollinate those with again.

6 SIXTH GENERATION
By the 5th or 6th year, your population should be so stable that you can grow the plants outside if no other calliopsis are growing nearby. Save seeds from the best plants.

Checklist
Season: Spring through autumn

Tools: Small watercolour brush

Equipment: Wire plant cage

Supplies: Horticultural fleece, strong packing tape, string, coloured electrical tape, plastic plant labels, waterproof marking pen

What can go wrong
Empty seeds: Counting on self-pollination from a flower that is self-incompatible can leave you with empty seeds. Guard against this by hand-pollinating unfamiliar plants.

Harvesting and cleaning

Producing seeds is only half the job when you're saving seeds, selecting improvements, or breeding plants. You also have to harvest the seed and clean and sort it for winter storage.

Harvesting seeds is one of the joys of gardening. But, as with any other gardening technique, you must prepare for it. Your harvesting technique will vary according to the plant you are growing. Say, for example, that you are growing just one type of sunflower (*Helianthus annus*) – an open-pollinated cultivar. To save its seeds, keep an eye on it as the seeds mature. Nature will tell you when it's ready – the minute the first bird investigates it, cut off the seedhead and finish drying it inside.

The problem comes with seeds that can shatter or fall to the soil when you try to pick them, or those in pods that pop, scattering the seed. To prevent losing the seeds, pollinate the flowers from which you plan to save seed, and then "bag" them, as shown opposite.

Cleaning

You will want to clean your seeds before storing them. As you'll see when you separate them from their seedheads or pods, bits of debris, or chaff, get mixed in. This doesn't hurt the seeds, but you'll prefer planting seeds that are relatively free of chaff.

If the seeds are medium size, it's easiest to put them in a shallow bowl and blow into the bowl – chaff will fly out. Wear goggles and a face mask when you do this!

You can put round seeds on a dinner plate. Hold the plate over a bowl, and tip it just enough to let the seeds fall into the bowl.

You can also use sieves to clean seeds. Commercially available sieves have screens of various sizes. You can use two sieves and a bottom, solid tray of some sort. The top sieve should have fairly large holes – it will catch the biggest pieces of debris but let the seeds and small pieces through. The next sieve should be the right size to catch the seeds – make sure that it's just slightly smaller than they are – but let the smaller pieces of chaff fall through to the sieve below. If you end up with too much seed-size chaff, you can blow it off, as described above.

Seeds of some fleshy fruit, such as tomatoes, have to ferment in order to become viable. To save these seeds, squeeze the seeds, the surrounding gel, and a little of the fruit into a bowl. Let the bowl sit in a warm – 18° to 24°C – location until a white mould appears on the top, and then pour it off. Put the seeds in a bowl and fill it with water. Empty seedcoats and debris will float to the top, and good seeds will sink to the bottom. Pour off the debris, and then pour the seeds and remaining water into a fine-meshed sieve to rinse them. Blot the bottom of the sieve with a kitchen towel, and then scrape the seeds onto a ceramic plate, a silicone baking mat, or a fine-meshed screen – they stick to paper as they dry. Spread them out thinly, and stir them repeatedly throughout the day. Don't put them in sunlight, which could overheat them. A fan that circulates air in the room can speed drying, but cover the plate to keep the seeds from blowing away.

Antirrhinum majus pods split when the seeds are mature, so it's wise to enclose stalks in bags when they begin to brown.

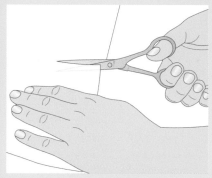

1 PREPARE
Cut a large enough piece of horticultural fleece to cover your seedhead.

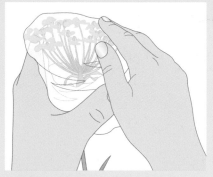

2 BAG
Fit the row cover material loosely over the seed head, leaving enough slack so pollen can blow around inside the bag.

3 FILL GAPS
Surround the stem with cotton wool or a thick layer of gauze to catch falling seeds.

4 LET DRY
As the stem withers and dries, the seed head will naturally droop to the ground and ripe seeds will fall into the bag.

5 CUT
When the stem seems nearly dry, cut it and carry it, with the bag still on it, into the house.

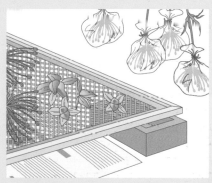

6 DRY
Let the seeds fall out of the seed head and dry for a few more days before you clean them and store them for the winter. Set aside part of an inside room, well away from the rain, snow and cold temperatures outside, where you can hang some plants to dry. Dry others on elevated screens.

Let white mould grow on the surface of the tomato gel and then rinse and dry the seeds as directed opposite.

What can go wrong
Seeds fall out of bag: Be certain that the bottom of the bag over your drying flower is completely plugged with gauze or cotton wool to prevent seeds from dropping to the ground.

Storing saved seeds

Saving seeds is fairly simple, but it does take time and thought. Give equal attention to storing your seeds – it's a shame to let them deteriorate over the winter. If you handle them well, you can keep most seeds for at least 6 years without a decline in germination rate. But it only takes a few days in poor conditions to kill them.

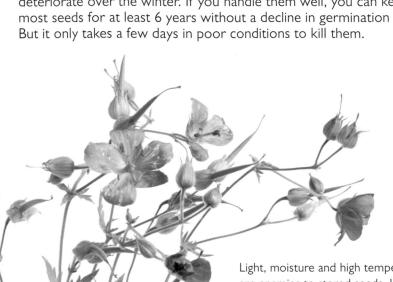

Most seeds are mature when they are brown and dry.

Geraniums can be grown from seed in soil with a consistent 10°C temperature.

Light, moisture and high temperatures are enemies to stored seeds. Light and high relative humidity can stimulate premature germination, high temperatures can bake seeds, and excess moisture can promote the growth of fungal diseases and rot.

Most seeds survive storage best if they contain about 8 per cent moisture. They can contain slightly less than this, but levels under 3 per cent kill them. Fortunately, there is a relatively simple way to dry them to the correct level, as shown opposite.

Once seeds are properly dry, you'll need to pack them. It's tempting to simply pour them into labelled paper envelopes and stick them in an old shoebox, but they don't keep well in those conditions. Properly dried and stored in an airtight container in the freezer, they will keep for years. But take care to dry seeds well so that excess moisture inside doesn't destroy them by expanding on freezing.

Some seeds must experience alternate thawing and chilling before they can break dormancy and

germinate. Good seed companies indicate this requirement, and it's also noted in the Plant Directory beginning on page 134 for plants covered in this book. Pack these seeds in a jar reserved for this purpose and make a note on the calendar. When the time comes, take the jar out of the freezer, let it thaw for a week or so, then freeze and thaw it again before planting.

If you live in a cold climate, you might wonder how to treat seeds that must be planted "when ripe", as noted in the Plant Directory or seed catalog. The answer is just that: plant them as soon as they come off the plant. But don't plant them in a nursery bed. Instead, use a seed tray. Fill the tray with a compost-based soil mix, water it, let it drain, plant it and cover it with thick plastic. Place it in a cold frame outside or in some other protected spot where it will be safe over the winter. In spring, move the tray into a good growing spot. Remove the plastic cover as soon as you see the first sign of germination.

Checklist

Equipment: Glass baking pan, glass jars, kitchen scale, seed trays

Supplies: Silica gel, compost-based soil mix, paper or horticultural fleece for seed packets, tape for packets

Store seeds in waterproof materials such as aluminium foil, waxed paper, or clingfilm, and tape the container shut to make a tight seal.

What can go wrong

Seeds rot: If the seal on your jar is not tight, moisture can get into the seed packets and rot them. Use duct tape to seal around the lid if you have any doubts about it.

Seeds dry out: Be sure to shake excess silica gel from the packages before you store them so it doesn't keep absorbing the moisture in the seeds.

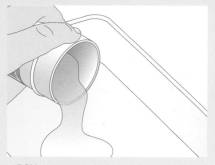

1 DRY
Pour silica gel in a 20cm x 20cm glass baking pan and heat it for 25 minutes on high in a microwave, stirring at intervals. This will dry the material.

2 PACK
Make a paper or horticultural fleece envelope and pour seeds into it. Tape it shut. Moisture can migrate through the paper or polyester as the seeds dry.

3 WEIGH
By placing an equal weight of silica gel and seeds into a jar, you'll dry the seeds to the correct moisture level.

4 SEAL
Leave the seed envelopes in the silica gel for 7 or 8 days. Test flat seeds by folding in half; if seeds break, they are dry enough. Test round seeds with a fingernail; the seedcoat should resist your nail.

5 STORE
Put seed packets in a sealed jar in the freezer to store them over the winter or even for a few years.

6 THAW
Let seeds come to room temperature before opening the sealed jar, and then allow seeds to sit for several days to adjust to ambient temperatures and humidity levels before planting them.

Treating seeds to prevent disease

A few plant diseases are soil-borne. If you are selling seed, you must treat seeds from disease-prone species to avoid passing trouble to your customers. On a home level, you may decide to treat only those seeds that came from plants you suspect of hosting an infection.

Some diseases are difficult to avoid. For example, by the end of the season, almost all tomato plants come down with an air- or seed-borne fungal or bacterial infection such as septoria leaf spot (*Septoria* spp.), geraniums and begonias develop bacterial spot (*Pseudomonas* spp.), and zinnias and petunias show symptoms of early blight (*Alternaria zinniae*).

Even if you don't intend to save the seed with the disease, it's sometimes unavoidable. The solution is to treat the seeds after removing them from winter storage when you are ready to

plant. All treatment techniques get the seeds wet, and they also condition them, or prepare them for planting.

Choices

You can treat seeds with hot water or with chemicals. Both systems require precision – you'll need to measure materials or temperatures accurately and time your treatments exactly.

Hot-water treatments, as shown opposite, are the kindest, easiest ways to treat seeds. They kill most common fungal and bacterial pathogens, although they don't kill viruses. But the chemical treatments don't reliably kill viruses either, so it's best not to save seed from a plant with a virus infection.

TREATING SEEDS AT HOME

Two chemicals are commonly used to treat seeds on a home scale.

Information about the dilution rates of these materials varies, depending on the source, as does the timing. To be safe, do not exceed the following times or concentrations, and to be doubly safe, treat no more than half of your saved seed. If this seed germinates without problems, you know that you can go on and treat the balance of the seed when you want to plant it.

• **Laundry bleach:** Dilute it by using from ½ to 1 part bleach to 9 parts water. Soak small seeds about 5 minutes; medium ones about 10 minutes; and large, tough ones about 15 minutes. Rinse immediately.
• **Food-grade hydrogen peroxide:** Dilute it by using from ½ part to 1½ parts hydrogen peroxide to 9 parts water. Soak small seeds for 5 minutes; medium ones about 10 minutes; and large, tough ones about 15 minutes. Rinse immediately.

Checklist

Season: Spring

Tools: Dairy thermometer, kitchen tongs

Equipment: Kitchen timer, ceramic or glass containers (for cooling)

Supplies: Cheesecloth, string, hydrogen peroxide, laundry bleach

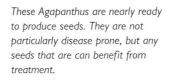

These Agapanthus are nearly ready to produce seeds. They are not particularly disease prone, but any seeds that are can benefit from treatment.

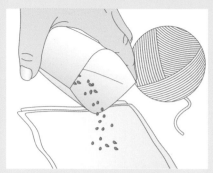

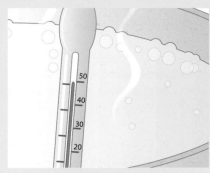

1 PREPARE

Make a "bouquet garni" of your seeds just as you do for spices when cooking. Triple-layer some cheesecloth and cut out a square that's large enough so that seeds will have ample room to move around in it. Tie it securely with cotton string.

2 MEASURE

Use a commercially available dairy thermometer rather than a sugar thermometer because it will be more accurate. Most seeds are sterilized at 50°C.

3 SUBMERGE

Wait until the temperature has held a stable temperature of 50°C for a few minutes before dropping the seeds into it. Begin timing immediately.

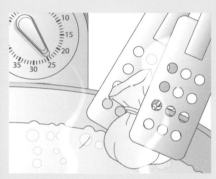

6 PLANT

Don't let the seeds dry out before you plant them – they will die if they do. Prepare the trays while the seeds are being sterilized and plant them as soon as you can handle them.

7 HEAT

Treated seeds germinate best if their soil is at the optimum germination temperature, so move the tray to a heating mat. Seeds will germinate more quickly than usual.

4 REMOVE

Research times and temperatures for vegetable seeds. For ornamentals, leave small seeds in the water for 20 minutes and larger seeds for 25 minutes. Remove with tongs.

5 LEAVE TO COOL

Let seeds cool enough so you don't burn yourself, but don't leave them too long.

What can go wrong

Seeds die: If chemicals are in too great a concentration, seeds are left in the solution too long, or seeds dry before they are planted, they will die.

Diseases strike: It could be that the treatment was not adequate to kill the disease or, equally likely, that the soil you planted into was also infected.

Secure the cheesecloth bundles tightly with string.

Working with hybrids

Working with hybrids – either to develop them or breed back from them to develop an open-pollinated cultivar – is easier than you might imagine. It requires careful record keeping, attentive care of your plants and patience over many years – but you won't need a degree in genetics to be successful.

Many Alcea rosea are open-pollinated plants that can be hybridized relatively easily.

Gardeners who are concerned about the loss of genetic diversity in the world sometimes place part of the blame on hybrids. As a consequence, they are very happy to learn how to "dehybridize" a hybridized cultivar, but they can't imagine taking the time to develop one. If you feel this way, remember that commercial hybrids often have increased vigour, yields and disease resistance, as well as other characteristics the breeder desired, such as a particular flower colour or size. When you hybridize, you try to create the qualities that you want in a plant; when you dehybridize, you try to stabilize the qualities you liked in a hybrid so that they breed true from open pollination.

The directions on these pages can get you started hybridizing and dehybridizing, but if you become fascinated by breeding your own plants, you'll want to research further.

APPROPRIATE PLANTS

Any open-pollinated plant is appropriate for hybridizing, and any hybrid plant is appropriate for dehybridizing.

Plants are classified as "inbreeders" if they can self-fertilize and "outbreeders" if they must be fertilized by pollen from another plant. In practical terms, it's easier to stabilize qualities in inbreeders than outbreeders because you can use a smaller number of plants – and, therefore, have less genetic diversity – to develop your cultivar. Inbreeders generally have small, inconspicuous flowers, while outbreeders have colourful, showy blooms to attract pollinators. If in any doubt about a plant, treat it as an outbreeder.

Checklist

Season: Spring to autumn

Equipment: Wire cages for plants, glass jars

Supplies: Seeds, tape, string, cotton wadding, labels, horticultural fleece, duct tape

What can go wrong

Plants are weak: After the first few generations, homebred plants may become weak and lack vigour. This is generally caused by a lack of genetic diversity within the strain. Avoid problems by growing a minimum of 20 – but preferably 30 – plants of your new cultivar each year and saving some seeds from each of them.

Plants get scrambled: Organization is the key to success. Label every packet of seeds and each planting with all possible details – parents, generation, dates, descriptions – and remember to keep separate plantings isolated from each other with geotextile or horticultural fleece cages.

Hybridize

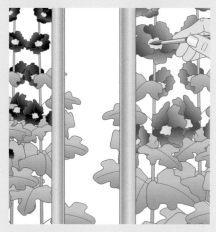

1 SAVE SEED

Isolate the two parent plants that you want to cross in cages. If you do this for a couple of years and save and replant seeds from these plants, you can develop relatively pure strains. If a plant such as a hollyhock becomes too tall for the cages, cut off the topmost growth; sideshoots will develop and flowers will form on them.

2 POLLINATE

Cross-pollinate the two species of parents by moving pollen from one type to the other. To be sure that you will be saving only cross-pollinated seeds, remove all other flowers from the plant as they form.

3 F2 GENERATION

Save seed from your first cross and plant it the following year. Cross-pollinate again and save seed. The next year, you'll have the F2 generation, and many of these will be showing the desired characteristics. Cross-pollinate and save selections again. Expect to do this for four or five generations to get a somewhat stable population.

Dehybridize

1 SAVE SEED

If you fall in love with a hybrid plant such as this cascading snapdragon (antirrhinum majus) and want to "dehybridize" it, begin by saving seed from it.

2 SELECT

The following year, grow at least 20, but preferably 30, plants from your saved seed. Keep only the plants that show the desired characteristic – in this case, a cascading habit – and throw out the rest before they bloom.

3 SELECT AGAIN

The following year, select plants again. Plant seeds in a tray on a heating mat or another area where the soil will maintain optimum temperatures and germinate more quickly. Eventually, after five or more generations, you'll have a fairly stable population. Remember to grow enough plants each year so that your gene pool doesn't become too restricted.

DIVIDING ESTABLISHED PLANTS

Dividing established plants is one of the easiest of all propagation techniques. It is also one of the safest – whenever you can choose between division and a more risky propagation technique, choose to divide. As long as you pay attention to a few basic techniques and guidelines, you'll never lose a plant or have a propagation failure as a result of dividing it.

Many herbaceous perennials need to be divided every few years to stay healthy. If you want to have more plants, you can divide them as soon as they are large enough to survive being split in half. Even if you don't want new plants, you must divide some of them within a year of spotting these symptoms.

Centre die-out: If clumps of multistemmed plants such as bergamot (*Monarda* spp.) become more vigorous on the perimeter than they are in the centre, you can assume that they need to be divided. This is sometimes described as doughnut growth, because the centre of the plant eventually dies back, and only the stems on the edges are alive.

Small blooms/few flowers: If a plant is growing in good soil but the blooms are smaller than normal, it probably needs dividing. Plants with unusually few flowers are also signalling distress and need dividing.

Sparse regrowth: If plants in good soil begin to put out fewer stems in the spring than normal, or fewer bottom leaves, they are also likely to need dividing.

Taking over the neighborhood: Some plants remain healthy despite outgrowing their allotted space. They can become so vigorous that they begin to shade out their neighbours or encroach on walkways and patios. Dividing these plants is the best way to keep them in bounds.

Above: Wood sorrel (Oxalis) plants are best divided in early spring before they begin vigorous growth.
Left: Bergamot (Monarda) lets you know when it needs dividing by dying out in the centre of the clump.

Good digging

Digging is an essential element of most kinds of division. Once you make digging a part of your gardening routine, you'll appreciate doing it in the most efficient way possible. Good digging technique not only saves your back, it also serves to protect the plant from unnecessary damage.

Prepare to dig your plants by thoroughly watering them at least 12 to 24 hours beforehand. Soak the area well, by either running a slow trickle from a hose near the plant's crown for a few hours or using a sprinkler in the morning or evening. Before turning off the water, sink a spade into a nearby spot to test whether the soil is truly saturated. Probe to at least the depth that you imagine the bottom roots grow. If the soil at the lowest level isn't wet, water until it is.

Plan to dig on a cool morning, preferably one with cloud cover, so the plant doesn't suffer while you are working. Begin by brushing off any mulch for a distance of 15cm to 20cm wider than the spread of the foliage. Dig with a spading fork or flat-bottomed spade rather than a shovel, and remember to sharpen your tools before you begin.

Sink the spade or fork to its full depth, working outside of the drip line of the plant's foliage. Lift the tool and sink it again. Continue in this way, circling around the plant, until you are back where you made your first cut into the soil. Remember to keep your back straight – to make this good posture natural, hold the tool handle with your thumbs pointing up to the sky rather than down to the ground.

Circle the plant again, pushing the tool even further into the soil if the roots extend beyond the first cut. When you begin your third circle, use the spade or fork like a lever to begin lifting the root ball out of the soil.

Most of the root balls you'll be digging will be small enough to simply lift out of the soil on the third circuit. However, if you are digging a very large plant, ask a friend to push a narrow wooden plank under the root ball as you work. The plank can then serve as a longer lever that will allow you to remove the root ball from the soil.

The roots must stay cool and moist until they are replanted. If you cannot immediately divide and replant, move the plant to a shady spot, and cover the roots with mulch or pieces of moistened hessian.

Above: Keep your back as straight as possible while you're digging.
Below: Retain as much of the root ball and smaller root hairs as you can.

Dividing plants with fibrous root systems

Root systems on mature plants are a wonder. They can be a dense, tangled mass that seems to grow in every direction, or a tidy collection of well-organized strands and delicate root hairs. No matter what kind of root system you're dealing with, you'll want to decide if you can gently pull it apart with your bare hands or if you'll have to use a tool to cut it into pieces.

leaves

stem

crown

Roots can be so dense that you must cut through them.

root system

If you examine a fibrous root system while the plant is dormant, you'll see buds on the crown at the top of the root system. Each of these buds is poised to become a new stem as soon as the plant begins spring growth. Theoretically, each of these stems and the accompanying roots is capable of becoming a new plant. However, single stems with tiny root systems are too vulnerable to plant; alone, they easily fall prey to a dreadful accident, a pest, or disease. Assure your success with divisions by dividing the parent root ball into no more than two or three pieces – this gives each section enough growth buds and roots to withstand normal garden conditions.

Timing is important when you are dividing plants. The general rule is to divide spring and summer bloomers in the autumn, and autumn bloomers in very early spring. However, if you live in a cold, short-season climate, this is impractical because autumn-planted divisions won't have enough time to become established before winter. In this case, make your divisions in the early spring. Wait until the ground has thawed enough to dig, but not until it's warm enough to encourage excessive underground growth.

What can go wrong

Division fails: You must treat the new division like a newly planted bare-root plant until it becomes established. Make certain that its soil remains consistently moist. If you notice wilting in moist soil, cut back the top growth or remove some leaves to reduce water needs. Once the roots become reestablished, the plant will be able to sustain normal growth.

Don't try to replant a division that has only a few roots.

1 PREPARE
Make the hole for transplanting the division into before you divide the plant. This allows you to replant the division right away.

2 DIG
If the soil isn't moist, water the evening before you dig. You'll want to get the entire root ball, not pieces, so take your time in working around the plant.

3 TIDY
The centres of many plants die out. When you divide them, you have an ideal opportunity to remove the dead growth and replant only the strong, new growth.

4 WATER
Divisions must not dry out before they are replanted. If you have to leave them, put them in the shade, cover them with hessian or newspaper, and water them well.

5 PLANT
When you plant the transplants, make certain to press down on the soil you are using to backfill the hole to remove air pockets. When the hole is halfway filled, water it well. Continue filling and pressing.

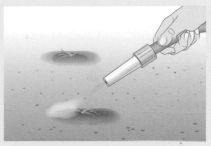

6 MAINTAIN
When you finish filling the hole, water the plant. Plants require steady moisture supplies while they are getting established. Water them consistently through their first season, and mulch them well once the ground freezes in early winter.

Checklist

Season: Early spring or autumn

Tools: Spade, spading fork, trowel, garden knife, small saw

Equipment: Watering can with rose nozzle, bucket of water

Supplies: Compost, soil improver, hessian

Temperature: Cool, but not windy

Humidity: Not important

Light: Cloudy

APPROPRIATE PLANTS

The following are a few of the plants with easily divided root systems. Many of these must be divided every few years to maintain their vitality, as shown by the column labeled "Frequency". Remember that the times given are averages – most of the plants in the particular genus require division at that frequency. Your plants could be growing more quickly or slowly, so make decisions about when to divide on an individual basis rather than by following this or any other recommendation.

COMMON	BOTANICAL	FREQUENCY	SEASON
Yarrows	*Achillea* spp.	2–3 yrs	Early spring or autumn
Sea Thrifts	*Armeria* spp.	3–5 yrs	Early spring
Asters	*Aster* spp.	1–3 yrs	Early spring
Bellflowers	*Campanula* spp.	3–5 yrs	Early spring or autumn
Chrysanthemums	*Chrysanthemum* spp.	1–2 yrs	Early spring
Tickseeds	*Coreopsis* spp.	1–2 yrs	Early spring or autumn
Purple Coneflowers	*Echinacea* spp.	4–5 yrs	Early spring or autumn
Blanket Flower	*Gaillardia* x *grandiflora*	3–5 yrs	Early spring
Geraniums	*Geranium* spp.	2–4 yrs	Early spring or autumn
Bee Balms	*Monarda* spp.	3 yrs	Early spring or autumn
Sundrops	*Oenothera* spp.	1–3 yrs	Early spring
Beard Tongues	*Penstemon* spp.	1–3 yrs	Early spring
Phlox	*Phlox paniculata*	3–4 yrs	Early spring or autumn
Black-Eyed Susans	*Rudbeckia* spp..	3–4 yrs	Early spring or autumn
Lamb's Ears	*Stachys byzantina*	2–3 yrs	Early spring or autumn
Speedwells	*Veronica* spp.	3–5 yrs	Early spring
Ornamental grasses	Various types	3–4 yrs	Early spring

Dividing plants with fleshy or woody crowns

crown

roots

The first time you dig up a hosta or canna, you might be surprised by the plant's crown. Rather than being small and loosely arranged, it's sizable, dense and packed with growth buds. Even though dividing these crowns and root systems might appear somewhat intimidating, you'll soon find that it's easy – as long as you have good tools.

After you dig up a plant that you want to divide, examine the crown, looking for the growth buds. Each of these buds has the potential to become a whole new plant, but like those on fibrous-rooted plants, single buds on a small piece of crown and with only a few little roots are vulnerable to all sorts of misfortune. To assure success, divide the crowns so that they include at least three or four growth buds and a sizable root system; two to three divisions is the usual number you can get at one time from a plant.

Some plants with fleshy or woody crowns are resistant to being divided. Peonies (*Paeonia* spp.) and bleeding heart (*Dicentra* spp.) lead this list. Cuttings are a much safer and more reliable propagation method. It's certainly possible to divide them if you absolutely must, but they will remain perfectly healthy if you don't ever dig them up.

Sharp tools are imperative for cutting through fleshy or woody crowns. Tools should be sharp enough to slice or cut through the crown without you exerting much pressure. In contrast, if you use a dull tool, you're likely to bruise the surrounding tissue as you struggle to cut through it. Buy good-quality garden knives and saws, and sharpen them before every use.

What can go wrong

Division dies: Several conditions can kill a division: it might be so small that it easily succumbs to a disease or pest problem; it could die of drought, as a consequence of having too few roots relative to the crown and emerging stems; nutrients in the new planting hole could be so concentrated that they burn tender root hairs. Protect divisions by cutting them with at least four growth buds and an adequate root system, and mix slow-release nutrient sources such as compost and rock powders into the surrounding soil before planting.

Checklist

Season: Early spring or autumn

Tools: Spade, spading fork, trowel, garden knife, small pruning saw, secateur (pruner)

Equipment: Plank of wood, watering can with rose nozzle, bucket of water

Supplies: Compost, soil improver, hessian

Temperature: Cool, not windy

Humidity: Not important

Light: Cloudy

APPROPRIATE PLANTS

Don't unthinkingly adhere to the following schedule. Let the condition of the plant tell you whether or not it needs, or can withstand, dividing.

COMMON	BOTANICAL	FREQUENCY	SEASON
Monkshoods	*Aconitum* spp.	Rarely	Early spring or autumn
Lilies of the Nile	*Agapanthus* spp.	Rarely	Early spring
Wormwoods	*Artemisia* spp.	3–5 yrs	Early spring or autumn
Goatsbeards	*Aruncus* spp.	3–5 yrs	Early spring or autumn
Astilbes	*Astilbe* spp.	3–5 yrs	Early spring or autumn
Cannas	*Canna* spp.	3–4 yrs	Early spring
Delphiniums	*Delphinium* spp.	Rarely	After blooming
Bleeding Hearts	*Dicentra* spp.	Rarely	Early spring
Hellebores	*Helleborus* spp.	3–5 yrs	Early spring
Daylilies	*Hemerocallis* spp.	3–6 yrs	After blooming
Hostas	*Hosta* spp.	Rarely	Early spring or autumn
Red-Hot Pokers	*Kniphofia* spp.	Rarely	Early spring or autumn
Lupines	*Lupinus* spp.	Rarely	After blooming
Peonies	*Paeonia* spp.	Rarely	Early spring or autumn

Paeonia withstand division if you're careful to injure their roots as little as possible.

1 DIG

Divide plants on a cloudy day. If you are dividing in autumn, do it at least a month before you expect a hard frost.

2 WASH

Wash off the soil around the roots because it's easier to make divisions if you can see the roots and crown of a plant.

3 CUT

It sometimes takes a garden saw to cut through a dense growth of roots. Make sure it's sharp before using it.

4 TRIM

Cut off all broken roots before you plant to help protect the plant from infection. If you're planting in autumn, trim back the top growth as well.

5 WATER

When replanting, remember to drive out air pockets from the soil and water the plant into its new position when the hole is half filled with soil.

6 MAINTAIN

Keep the plants consistently moist. After the top of the soil has frozen in early winter, mulch well to prevent thawing and heaving for the next few months.

Dividing suckers

Suckers form at nodes along a plant's underground stems – the stolons or rhizomes. You may have noticed the suckers of raspberries, lilacs and forsythia in your garden and may even have tried to weed them out of areas they were invading. They can be difficult to discourage, but if you sever the sucker from the parent plant to create a new individual, you'll not only get rid of the invader, you'll also have a plant to share.

Many woody plants form suckers once they have grown to their mature height, probably because they are an excellent way to assure a plant's survival. The suckers that form on the perimeter of the parent plant are capable of becoming independent once they have a good root system. If space is unlimited, many suckers can grow around the original plant without causing it problems. However, in restricted areas or where the soil is poor, it's wise to limit the number of suckers that compete with the parent plant.

On most plants, you can divide the suckers that formed the previous summer and transplant them without a problem. However, if you are at all uncertain about the sucker's ability to survive on its own, prepare it a year before dividing it. In early spring, explore the sucker to find the stolon from which it's growing. Sink your spade in a circle around the sucker, about 20cm to 30cm away from its stem, on all sides except through the stolon. Repeat this several times throughout the season, each time being careful to spare the stolon. The sucker will respond by concentrating its new roots inside the "pot" you've created.

Checklist

Season: Early spring

Tools: Spading fork, garden knife or saw

Equipment: 30cm pot, watering can or hose with rose nozzle or water breaker

Supplies: Hessian, compost, soil improver, potting soil

Temperature: Cool, but not windy

Humidity: Not important

Light: Cloudy

leaves

stems

sucker

good-size root ball

You can see the suckers to the left of the main stem.

1 OBSERVE
In order to divide a sucker, you must first find the stolon on which it is growing. Explore gently so you don't hurt the parent plant.

2 EXAMINE
Before you decide to cut a sucker from the parent plant, check to see that its root system is large enough to sustain the plant.

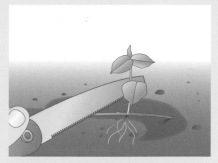

3 CUT
As always, you will want to use the best tool to cut the sucker from the plant. In many cases, this is a small, sharp garden saw because it is so strong. While you are cutting the sucker from the parent plant, be careful not to injure any more roots than is necessary.

4 REPLANT
If you already have too much of a particular plant, pot the suckers and give them to friends. Wrap unplanted roots in wet hessian.

5 MAINTAIN
Keep the plant consistently moist, and encourage it to grow roots by picking off any flower buds that form the first year.

Forsythia plants are easy to propagate from suckers or by tip layering.

What can go wrong
Divided sucker dies: A very small sucker may not have the vitality to become established before the weather gets hot and the plant needs more water than the root system can supply. Cut back the top growth if it seems too large for the root system.

The sucker at left does not have enough roots to plant.

APPROPRIATE PLANTS

COMMON	BOTANICAL	SEASON
Bottlebrush Buckeye	*Aesculus parviflora*	Spring
Serviceberry, Juneberries	*Amelanchier* spp.	Spring
Chokeberries	*Aronia* spp.	Spring
Carolina Allspices	*Calycanthus* spp.	Spring
Summersweets	*Clethra* spp.	Spring
Shrub Dogwoods	*Cornus* spp.	Spring
Forsythias	*Forsythia* spp.	Spring
Shallons	*Gaultheria shallon*	Spring
Smooth Hydrangea	*Hydrangea arborescens*	Spring
Kerria	*Kerria japonica*	Spring
Bayberries	*Myrica* spp.	Spring
Pachysandras	*Pachysandra* spp.	Spring
Poplars	*Populus* spp.	Spring
Rugosa Rose	*Rosa rugosa*	Spring
Elderberries	*Sambucus* spp.	Spring
Snake Plants	*Sansevieria* spp.	Spring
Lilacs	*Syringa* spp.	Spring

Dividing plants with rhizomes

When you think of rhizomes, you probably think of the bearded iris with its prominent rhizomes emerging from the ground. But as you can see from the list at right, this certainly isn't the only garden plant that grows and can be propagated from a rhizome.

Timing matters when you are dividing rhizomes. In most regions it's best to wait until after the plant has finished blooming to dig up the rhizome and divide it. But if you live in a cold, short-season area, this may be impractical because divisions planted in late summer and early autumn don't have time to become established before the ground freezes. In this case, divide in spring and snap off any flower buds that form so the plant can put its energy into creating a strong root system for a year before it blooms.

You may wonder how often to divide rhizomatous plants. Some of them suffer if their rhizomes are allowed to grow unchecked, so dividing them every 3 to 5 years keeps them vital and strong. Once again, rather than adhering to a predetermined schedule, watch the plant and divide it when flowering decreases, flower size diminishes, or it just looks crowded.

APPROPRIATE PLANTS

COMMON	BOTANICAL
Cupid's Bowers	*Achimenes* spp.
Baneberries	*Actaea* spp.
Maidenhair Ferns	*Adiantum* spp.
Elephant's Ears	*Alocasia* spp.
Peruvian Lilies	*Alstroemeria* spp.
Wild Gingers	*Asarum* spp.
Blackberry Lily	*Belamcanda chinensis*
Bergenias	*Bergenia* spp.
Cuckoo Flower	*Cardamine pratensis*
Bearded Irises	*Iris* spp.
Lotus	*Nelumbo* spp.
Mayapples	*Podophyllum* spp.
Tuberose	*Polianthes tuberosa*
Solomon's Seals	*Polygonatum* spp.
Wall Polypody,	
Rockcap Ferns	*Polypodium* spp.
Christmas Ferns,	
Sword Ferns	*Polystichum* spp.
Bloodroot	*Sanguinaria canadensis*
Wakerobins,	
Wood Lilies	*Trillium* spp.

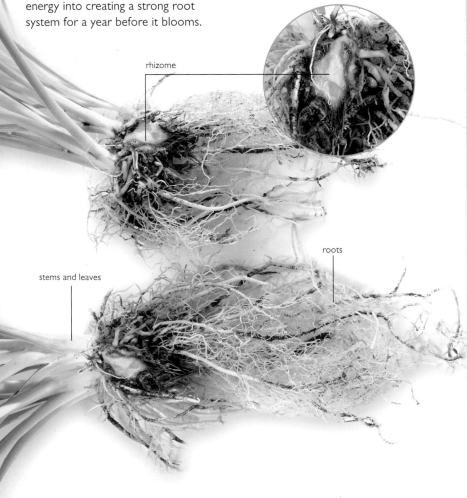

rhizome

stems and leaves

roots

What can go wrong

Rhizome doesn't grow: If you plant a rhizome with no buds, it cannot grow leaves and will die. If you are cutting a rhizome before the leaves have emerged, check to make certain that each piece has at least one, but preferably two, strong growth buds.

This rhizome should have more growth buds.

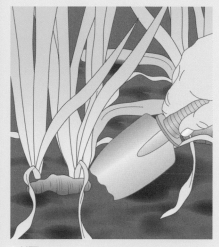

1 LIFT

Dig up the entire rhizome of the plant you want to divide, trying not to damage the root system.

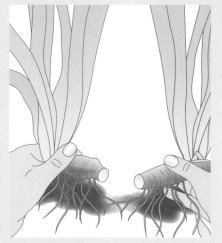

2 EXAMINE

Divisions will reestablish best if they have at least two growth buds with leaves growing from them as well as many roots.

3 CUT

Use a very sharp knife to cut the divisions. A blunt knife can bruise tissues because you have to press so hard on it.

Checklist

Season: Spring or after plants finish blooming

Tools: Garden fork, trowel, garden knife, pruners

Equipment: Garden hose, cutting board

Supplies: Compost, soil improver

Temperature: Cool, not windy

Humidity: Not important

Light: Cloudy

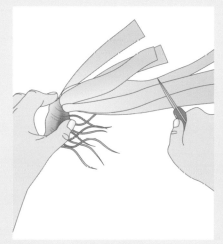

4 TRIM

Cut back the leaves so that the plant isn't stressed by having to provide water for a large leaf area.

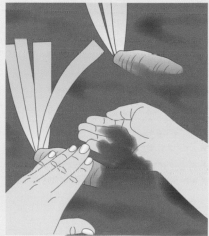

5 PLANT

Replant the rhizome at the same depth it was growing, and water it well to exclude air pockets from around the roots.

Propagate Bergenia by dividing rhizomes.

Dividing plants with tubers

Technically, tubers are swollen, underground stems – but so are rhizomes. Gardeners tend to distinguish between the two by the way they look (tubers have smoother skin and are more regularly shaped) as well as their depth under the ground (tubers tend to grow at a lower level). But these differences aren't worth worrying about. Whether a plant is said to have a tuber or a rhizome, it's the same sort of structure, and you'll propagate it in the same way.

When you think of tubers, you probably think first of potatoes. Beyond that, you're likely to think of dahlias or tuberous begonias. But just to add to the confusion, dahlias don't have tubers; their swollen, underground structures are actually roots, not stems. Fortunately, this distinction doesn't matter in a practical sense, either. A tuberous root is as capable of becoming a new plant as a tuber or rhizome and is planted in the same way.

Many tuberous plants come from warm regions where the soil doesn't freeze. As a consequence, their tubers can't survive a winter in freezing soil. To care for them properly, dig them up at the end of the summer and store them in a dark area that maintains temperatures around 4°C, then replant after the frost-free date the following season. Store them in sand or peat moss, and check them through the winter to see that they are neither drying out nor rotting. Add a sprinkle of water to the medium if they are drying out, and close the storage container tightly to retain this moisture. If they are rotting, throw away diseased tubers and let the remaining ones dry for a day or so before packing them in fresh sand or peat moss.

Stored tubers should be neither too moist nor too dry.

roots

growth buds

tuber

stem

What can go wrong
Tuber dies in storage: If you lift tubers before they have had a chance to develop thick, strong skins, they will be vulnerable to rotting fungi over the winter months. Similarly, if their storage medium is too moist, they may rot. If the medium is too dry, it may steal moisture from the tuber – check every few weeks, and adjust moisture levels as necessary.

Tuber rots in the ground: If you replant tubers in poorly drained soil right before a prolonged period of rainy weather, they will be vulnerable

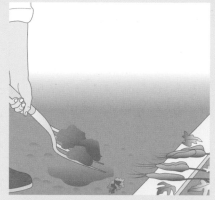

1 DIG

Let tubers rest for a week after the top growth has been killed by frost or you have cut it. This allows the skin to harden. Dig at least a foot away from the stems on all sides to avoid injuring the tubers.

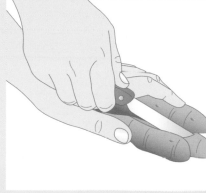

2 DRY

Let the soil around the tubers dry out before you handle them.

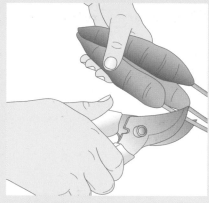

3 CUT BACK

Cut back the stems to several inches above the tubers. Next year's eyes will grow from the same area, so it's also important to check for diseases or broken necks. If you find any, discard the tuber.

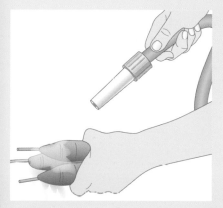

4 WASH

Gently wash the tubers again to get all of the soil off. Let their surfaces dry completely.

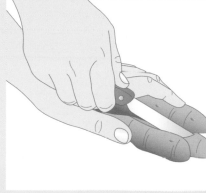

5 CUT

Cut the tubers apart so that you can grow them without crowding the next year. Make sure each tuber has at least one eye. If not, leave it attached to a tuber that does.

6 PACK

Dry the tubers overnight and then pack them, upside down, on top of a layer of sand or peat moss in a sturdy box.

to soilborne fungal diseases that cause rotting. Mix extra drainage material such as fully finished compost or even sand into the soil before planting.

Without an eye, this tuber won't be able to grow.

APPROPRIATE PLANTS

COMMON	BOTANICAL
Lords and Ladies	*Arum* spp.
Asparagus Fern	*Asparagus densiflorus*
Begonia, tuberous	*Begonia* x *tuberhybrida*
Caladiums	*Caladium* spp.
Hearts on a String, Fountain Flowers	*Ceropegia* spp.
Dahlias	*Dahlia* spp. & cvs.
Foxtail Lilies	*Eremurus* spp.
Gloriosas	*Gloriosa* spp.
Hellebores	*Helleborus* spp.
Daylilies	*Hemerocallis* spp.
Shamrocks	*Oxalis* spp.
Ranunculus	*Ranunculus* spp.
Calla Lilies	*Zantedeschia* spp.
Rain Lilies	*Zephyranthes* spp.

Dividing plants with offsets

Offsets are young plants that grow from the crown or a stolon of a parent plant. As you might imagine, distinguishing between an offset and a sucker can sometimes be difficult. But don't let the name of the young plant distract you. Even though an offset is not a sucker, you'll be severing it from its parent plant in much the same way you sever suckers.

Many of the plants that produce offsets come from warm, dry regions where the survival of germinating seeds is by no means guaranteed. Offsets – which take their water and nutrients from the parent plant until they are safely rooted on their own – are a safer means of propagation. You may find that the majority of plants you grow this way are houseplants and, consequently, you'll be moving them into pots rather than a garden bed.

APPROPRIATE PLANTS

Technically, bulbs such as daffodils are offsets. However, we are using the gardeners' convention of discussing them separately (see pages 68–73).

COMMON	BOTANICAL
Urn Plants	Aechmea spp.
Agaves	Agave spp.
Chinese Evergreen	Aglaonema modestum
Aloes	Aloe spp.
Pineapples	Ananas spp.
Flamingo Flowers	Anthurium spp.
Deer Fern	Blechnum spicant
Bromeliads	Bromelia spp.
Spider Plants	Chlorophytum spp.
Clivias	Clivia spp.
Swamp Lilies, Poison Bulbs	Crinum spp.
Sago Palms	Cycas spp.
Dyckias	Dyckia spp.
Echinopsises	Echinopsis spp.
Strawberries	Fragaria spp.
Haworthias	Haworthia spp.
Bananas	Musa spp.
Ortegocactus	Ortegocactus macdougallii
Orthophytums	Orthophytum spp.
Strawberry Begonia	Saxifraga stolonifera
Hen and Chicks, Houseleeks	Sempervivum spp.
Air Plants	Tillandsia spp.
Yuccas	Yucca spp.

leaves

stems

roots

This plant has produced offsets in all directions.

If you are going to replant an offset in the garden, think about your timing. In frost-free areas, you can remove and replant offsets in autumn and winter, but if these are your dry seasons, remember to keep them well watered while they are becoming established. In a cold-winter area, you'll want to be certain that an offset has 4 to 6 weeks to establish its root system prior to freezing weather. If you can't be sure of this, wait until the following spring to remove the new plant.

The maturity of an offset also determines your timing. Offsets that do not yet have a good root system can't live as individuals and will die. Rather than dividing an offset too soon, wait until the following year to sever it from the parent plant.

Checklist

Season: Spring or autumn

Tools: Garden fork, trowel, garden knife, pruners

Equipment: Pots

Supplies: Potting soil

Temperature: Cool

Humidity: Not important

Light: Not bright sunshine

What can go wrong

Offsets die: If they do not get adequate moisture, either because their root system is not yet large enough to supply the top growth with enough water or because their soil dries out, offsets can die. Only take offsets with healthy root systems, and keep their media consistently moist until they are established.

With too few roots, this offset will likely die.

1 LOOK
Examine the plant to find offsets that are large enough to live on their own.

2 UNPOT
Remove the plant from its pot. Don't try to dig out just the offset because you could injure the remaining roots, leaving the way open for disease.

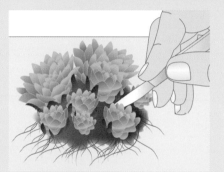

3 TEASE
Gently tease the offset, along with its roots, away from the parent plant.

4 PLANT
Plant the offset in its own pot as soon as possible – you don't want to let it dry out.

5 MONITOR
Place the plant in a spot with indirect light, and let it grow there until it's fully established in its new circumstances.

Wait until Chlorophytum plantlets have developed roots before severing them from the parent plant.

Dividing plants with bulbs and corms

Bulbs and corms are underground storage structures holding nutrients that the plant uses after it breaks winter dormancy and begins its seasonal growth. The cardinal rule when growing a plant with a bulb or corm is to refrain from cutting off its leaves after it has finished blooming. Let them die down on their own so they have time to manufacture the nutrients that will eventually sustain new growth.

Checklist

Season: Autumn

Tools: Spading fork, trowel

Equipment: Paper bag or cardboard box

Supplies: Peat moss or sand, water, plant labels

Temperature: Not freezing

Humidity: As dry as possible

Light: Not important

True bulbs include lilies, tulips, daffodils, snowdrops, grape hyacinths, *Scilla*, Dutch irises, and *Iris reticulata*. Gladiolus, crocus, autumn crocus, and freesia are corms. This distinction is important only because it makes a difference to how the plants grow.

Bulbs – which are actually modified flower buds and stems enclosed in scales or leaf bases – create new bulbs, called bulblets, around the perimeter of their basal plates or at their leaf bases. Eventually some of these enlarge to the same size as the original bulb and split off from it to form a new individual. You often see this when you buy daffodils – bulbs with a "double nose" are on their way to becoming two individuals. Thanks to the many maturing bulbs and bulblets, a bulb bed that's left undisturbed for a number of years becomes crowded. To keep your beds looking their best, you'll need to dig and divide them every 3 or 4 years.

Corms reproduce themselves each year. After the original corm has produced roots and top growth, a new corm grows to take its place. The plant may also produce tiny cormels. The corms and cormels of many plants are too tender to survive freezing soil, so you'll have to dig and store them through the winter months as you do tuberous begonias (see page 64).

Tulip bulb

basal plate

Daffodil bulb

bulblet

bulblet

basal plate

bulb

root

What can go wrong

Plants don't bloom: It may take 2 to 4 years for a small bulblet or tiny corm to grow to blooming size. If it's a tender corm, remember to lift and store it along with the rest of your corms each year. Your patience will be rewarded.

Corms or bulbs die in storage: Check your bulbs and corms every few weeks to make certain that they are not rotting, as a result of excessive moisture, or drying out. If they are too moist, take them out of the medium, let their surfaces dry, and repackage them in somewhat drier material. If they look a little shrivelled, sprinkle their medium with water and enclose them in a plastic bag for a few days to let them rehydrate. Remember to move them to a paper container in a few days so they don't suffer from excess moisture.

Dividing bulbs

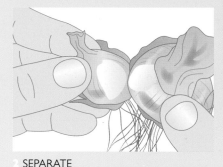

1 SEARCH
Sift through the soil with your fingers to find tiny bulblets that may have fallen off the parent bulb.

2 SEPARATE
Gently pull apart new bulbs once they have developed skin between the two bulbs; before this, the new bulb is too young to be divided.

3 PLANT
Plant the divisions the same way that you plant those you buy. If they are significantly smaller, plant them just a little bit less deeply than you plant full-size bulbs.

Dividing corms

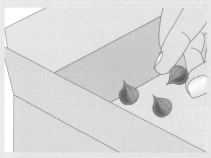

1 DIVIDE
In autumn, when you lift the corms, cut the stems off the corms and let their necks dry.

2 DISCARD
After discarding the old corm, pick off the baby cormels that may have formed around the base of the new corm.

3 STORE
Store both the new corms and the cormels in a labeled cardboard box or paper bag over the winter. Do not let them freeze. Plant them once the frost-free date has passed in spring.

Cover the planted bulblets with a scant 2cm of soil.

APPROPRIATE PLANTS

COMMON	BOTANICAL	SEASON
Ornamental Onions	*Allium* spp.	Autumn
Quamashes	*Camassia* spp.	Autumn
Giant Lilies	*Cardiocrinum* spp.	Autumn
Glories-of-the-Snow	*Chionodoxa* spp.	Autumn
Autumn Crocuses	*Colchicum* spp.	Midsummer
Crocuses	*Crocus* spp.	Autumn
Trout Lilies	*Erythronium* spp.	Autumn
Freesias	*Freesia* spp.	Autumn
Fritillarias	*Fritillaria* spp.	Autumn
Snowdrops	*Galanthus* spp.	Autumn
Gladioli	*Gladiolus x hortulanus* cvs.	Early summer
Bluebells	*Hyacinthoides* spp.	Autumn
Hyacinths	*Hyacinthus orientalis* cvs.	Autumn
Dutch Irises	*Iris* spp.	Autumn
Lilies	*Lilium* spp.	Autumn
Magic Lilies	*Lycoris* spp.	Summer
Grape Hyacinths	*Muscari* spp.	Autumn
Daffodils	*Narcissus* spp.	Autumn
Stars of Bethlehem	*Ornithogalum* spp.	Autumn
Squills	*Scilla* spp.	Autumn
Harlequin Flowers	*Sparaxis* spp.	Autumn
Tulips	*Tulipa* spp.	Autumn

Scoring, sectioning, scooping and coring bulbs

Although you may never use any of the following techniques, they can be invaluable if you come across a bulb you really love that's no longer available for sale. If it's a daffodil, you can simply wait until it creates bulblets in its normal fashion. But if it's a hyacinth, it may not. These plants are resistant to creating bulblets, so it's useful to learn a technique that stimulates their production.

Bulbs are classified as tunicate or nontunicate, depending on how their layers of leaf bases are arranged. If the layers grow in concentric circles – as in onions, tulips and daffodils – they are tunicate bulbs. In contrast, bulbs such as lilies are nontunicate because the bases of their leaves form overlapping layers, or scales. This distinction determines how you force the bulb to create more bulblets. You divide nontunicate bulbs into scales (as on pages 72–73) and use the techniques on this page with tunicate bulbs.

The bottom of the bulb, to which the leaf bases of both tunicate and nontunicate bulbs are attached, is known as the basal plate. It looks like a thick callus of some sort, but it's actually a very short stem with the capacity to grow roots as well as new leaves. Some of these techniques require its removal, while others call for injuring it enough to stimulate new growth.

<table>
<tr><td colspan="2">APPROPRIATE PLANTS</td></tr>
<tr><td>COMMON</td><td>BOTANICAL</td></tr>
<tr><td>Ornamental Onions</td><td>Allium spp.</td></tr>
<tr><td>Snowdrops</td><td>Galanthus spp.</td></tr>
<tr><td>Amaryllises</td><td>Hippeastrum spp.</td></tr>
<tr><td>Hyacinths</td><td>Hyacinthus orientalis cvs.</td></tr>
<tr><td>Daffodils</td><td>Narcissus spp.</td></tr>
<tr><td>Squills</td><td>Scilla spp.</td></tr>
<tr><td>Tulips</td><td>Tulipa spp.</td></tr>
</table>

Checklist

Season: Early summer

Tools: Garden knife

Supplies: Sand, vermiculite, or perlite

Temperature: Depends on technique chosen, but no cooler than 21°C

Humidity: Depends on technique chosen, but no drier than 80 per cent

Light: Place bulbs in darkness while they are developing bulblets.

Scored bulb

bulblets

cuts

basal plate

bulblets

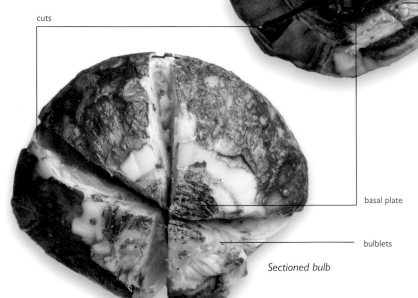

Sectioned bulb

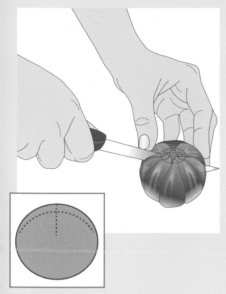

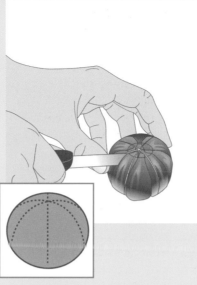

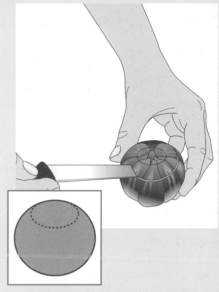

SCORING

Cut wedges into the bulb, through the basal plate and just beyond the widest circumference. Set the bulbs upside down, with the basal plate pointing upward, on a bed of fresh, clean sand, perlite, or vermiculite. Place the bulbs in a dark spot with warm temperatures (21° to 27°C) and high humidity (80 to 90 per cent) until autumn. At least 10 (but up to 20) bulblets will form by autumn. Plant the whole bulb, without dividing the bulblets, as you normally do in the garden. Mark the spot, and the following autumn dig up the bulb, divide the bulblets, and plant them in a nursery bed. They normally take 3 or 4 years to reach blooming size.

SECTIONING

To section a bulb, cut across it three or four times, making 6 or 8 pie-shaped wedges. Set the wedges upside down, with the basal plate pointing upward, on a bed of fresh, clean sand, perlite, or vermiculite, and place them in a dark spot with warm temperatures (21° to 27°C) and high humidity (80 to 90 per cent) until autumn. Bulblets will form at the basal plate of each section. Plant the wedges, without dividing the bulblets, as you normally do in the garden. Mark the spot, and the following autumn dig up each section, divide the bulblets, and plant them in a nursery bed. They normally take 3 to 4 years to bloom.

SCOOPING

Scoop bulbs by using a knife to remove the basal plate. Cut deeply enough into the centre of the bulb to remove the shoot and flower bud. Bulblets will grow from the leaf bases that are now exposed. Place the bulbs, upside down and on a clean medium, in a place with a 21°C temperature and relatively low humidity – 65 to 75 per cent – for about 2 weeks. The cut flesh will form a callus during this time. Check the bulbs daily during the third week, and when you see definite swelling, increase the temperature to 27°C and the humidity to 85 per cent. Continue to watch the bulblets, and when they form roots, plant them in a soil mix that contains nutrients. In autumn, move them to an outdoor nursery bed and let them grow, undisturbed, for 4 years. Move them into the display beds at this point, because they will be ready to bloom.

What can go wrong

Bulb rots: Sterilize your knife before cutting the bulb; place it in new, clean media or a clean cardboard box; and maintain the recommended environmental conditions to decrease the possibility of fungal diseases.

Watch out for rot.

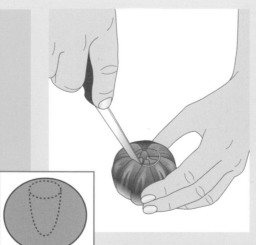

CORING

Core bulbs by cutting out a cone from the basal plate to the centre of the bulb, removing the stem and flower buds in the process. Place the cored bulbs, upside down and on a clean medium, in a dark spot with warm temperatures (21° to 27°C) and high humidity – 80 to 90 per cent – until autumn, and then plant them in the garden. Fewer bulblets will form with this method, but they will be large enough to reach blooming size in only 2 to 3 years.

Scaling lily bulbs

What could be more spectacular than a broad swath of lilies covering a hillside or bordering a long drive? If you're thrilled by the idea of this kind of display but can't imagine being able to afford it, you'll be delighted to learn how to propagate lilies by scales.

Unlike other bulbs, the lily bulb is formed of overlapping scales that detach from the basal plate without damaging the mother bulb. Each scale is capable of producing up to four bulblets at its base. Thus, you can radically increase your lily population at very little cost. But be forewarned, the young plants may take 3 or 4 years to reach blooming size, so it's best to plant a noninvasive ground cover with which they'll eventually share the area, while you're waiting for them to mature into a flowering display.

How long?
After the young lily bulbs have been transplanted to the nursery beds, they will grow quickly. A few may bloom the following season, but most will wait until their third year to produce flowers. Transplant them to their permanent spots a year or two after you put them in the nursery beds.

Another method
Rather than rooting the scales in a container, some gardeners place them in a plastic bag filled with moist vermiculite and set the bag in the proper environment. After they have formed roots and bulblets, plant them in individual 8cm pots and grow on, as described in the final step at right.

Checklist
Season: Spring

Tools: Trowel or fork (to dig the bulbs), your fingers (to pull off the scales)

Medium: Sterile, soil-less, moisture-retentive medium, such as peat-lite or half-peat and half-sand

Depth: Bury two-thirds of each scale in the medium.

Temperature: Medium should maintain 16° to 21°C temperatures.

Humidity: As long as the soil mix is moist but not soggy, there is no need to increase humidity levels.

Light: Indirect and diffused. Place the rooting tray in a somewhat shaded location to prevent the scales from overheating.

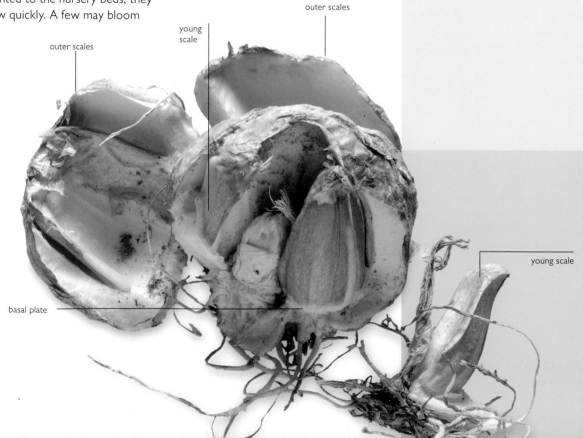

outer scales

young scale

outer scales

young scale

basal plate

1 CLEAN THE BULBS

After digging your bulbs or buying them, gently wash off all the soil. Not only will this allow you to see all the bulb parts, it also eliminates a potential source of fungal spores. Check for diseases, too, and discard any suspect bulbs.

2 REMOVE OLD SCALES

Gently pull on the outermost scales to remove them. The oldest scales won't be viable; the plant sloughs them off as it grows. But if the bulb is newly purchased, the oldest scales were probably removed before it was shipped, so these may be good.

3 DETACH VIABLE SCALES

To remove viable scales, gently pull each scale to the side and down so that it naturally breaks off where it joins the basal plate. This may seem alarming, so experiment until you develop a feeling for the correct angles and strength with which to pull.

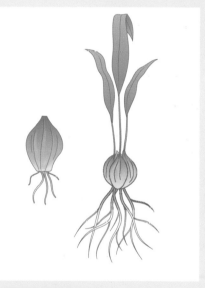

4 DIP AND DRY

Dip the scales in topical hydrogen peroxide or copper fungicide to act as a fungicide. Let the scales dry on a paper towel and wait until a protective callus forms on the base of the scale before planting it. This callus prevents many fungi from entering the wound.

5 DIBBLE THE MEDIUM

Make slots in which to place the scales. Leave 2½cm between them in every direction, and gently firm the medium around them. Set the rooting tray in an area with diffused light and temperatures of 16° to 21°C. Check the medium daily for moisture; water when dry.

6 GROW THE PLANTS

After several weeks, roots and bulblets will grow. When the first leaves are about 2cm tall, transplant each scale to a soil mix in an 8cm pot. Let them develop for a year; separate individual plants the following spring and plant in a nursery bed.

What can go wrong?

Fungal disease: Fungal infections can strike lily scales. Prevent problems by inspecting the stock bulbs carefully for disease, and discard any that look suspect. After pulling off the scales, dip them into a solution of topical hydrogen peroxide and let them dry on a paper towel. Plant them only when they have dried and you can see a slight callus on the broken edge. Water only enough to keep the medium moist.

Be especially vigilant if you are starting off more than one bulb to a pot. Disease can spread quickly from one plant to another.

CUTTINGS

Taking cuttings and stimulating them to root can be a simple, straightforward way to propagate your plants. In the case of tender, tropical houseplants and many woody vines and shrubs, it's frequently the recommended method. Not only does it give the surest results, it also guarantees that the new plant will be identical to the parent. However, to achieve success, you do have to follow some commonsense guidelines.

Guidelines for success

• Cuttings have the greatest chance of developing roots if you take them from plants that are relatively young and growing vigorously. Where you want to take a cutting from an old plant to preserve it, try to prepare the plant a year in advance. Prune back to old wood on a few stems, and give the plant balanced nutrition and an appropriate amount of water for a season. This treatment should stimulate healthy new growth the following spring.

• Always take cuttings from the current season's growth – even if you're taking hardwood cuttings in autumn, the wood must have grown that season, not the previous year.

• Take cuttings early in the morning, while the stem is still full of water. Place all cuttings (except hardwood) in water, or wrap them in damp paper towels in a plastic bag until you prepare them for rooting.

• Take cuttings only from pest- and disease-free plants. If you take a cutting from a plant with a pest or disease problem, it is likely to die rather than root. Cure problems before propagating a plant.

• Just before every cut, clean your knife by carefully wiping the blade with a cloth soaked in a 10 per cent laundry bleach solution (1 part bleach to 9 parts water).

• Take cuttings only from plants in excellent health. If a plant has been growing in less than ideal conditions, it may, for example, have responded by putting on soft, sappy growth or developing excessively long internodes. Such symptoms will make the cutting more susceptible to fungal diseases – the bane of almost all propagation methods.

• In the case of plants such as ivy that have both juvenile and adult foliage, take cuttings only from

Use the sharpest blade possible to take cuttings so you don't bruise the stem.

wood that produces the juvenile leaves – this is still in an active vegetative growth phase, whereas the wood with adult leaves is not.

• On all but hardwood cuttings, cut the base of the cutting on a slant to expose more of the interior to the rooting medium. On hardwood cuttings, use a slanted cut at the top and a straight cut at the bottom, so that you can distinguish between the top and bottom of the stem.

• Learn which sort of cutting to take from each plant species. Check the Plant Directory (see page 134) and other resources for this information.

• Take more cuttings than you will eventually need, but do not take so many that the parent plant suffers. If every cutting roots and you end up with too many plants, give them away to friends and neighbours.

Types of cuttings

When you think of cuttings, you're likely to think of the tender tips of houseplants that you cut and put in water to root. But the world of cuttings extends far beyond that. You'll learn to take cuttings of stems at all stages of growth, cuttings from leaves, and even root cuttings.

Conifers can be propagated from hardwood cuttings.

As you'll see in the following pages, there are various sorts of cuttings of different ages and different plant parts. Depending on the species, you can successfully grow a new plant from a piece of the stem, a leaf, or a root.

• Stem cuttings are categorized according to their age and include softwood, greenwood, semiripe and hardwood cuttings.

• When working with hardwood, you can take a straight cutting, a heel cutting, or a mallet cutting.

• Tropical plants such as dumb cane (*Dieffenbachia* spp.) have a central stem, or cane, that never becomes woody. Stem cuttings from these plants are known as cane cuttings and are handled differently than other types of stem cuttings.

• Leaf cuttings can be categorized as being leaf petiole cuttings, vein cuttings, upright leaf cuttings and monocot leaf cuttings.

• Root cuttings are simply that – pieces of root that are capable of growing both new top growth and a fully formed root system.

Meristematic tissue

Cuttings develop into new plants because they include some meristematic tissue, or cells that easily become meristematic tissue. Cells in this tissue are capable of dividing to produce new cells. Depending on the location of a meristem, it will produce roots or stems, leaves, and flowers.

In the root, the apical meristem is at the root tips. This location allows it to create cells that become the protective root cap, as well as those that form the body of the root.

In the stem, the meristem located at the tip of each shoot is also called the apical meristem. Meristematic tissue found in the axil of each leaf is known as the axillary meristem. Every bud in every leaf axil of most plants has the capacity to become a branch. However, because of a process known as apical dominance, many of them remain latent, or inactive. This is because the stem tip produces plant auxins, compounds that are similar to hormones and suppress the growth of secondary stems. If you pinch off the tip of the stem – as you often do when you pinch or prune – you remove these auxins. In response, the topmost leaf nodes begin to develop. Before

long, they produce their own auxins to suppress the growth of nodes below them on the stem.

Woody plants have an additional site for meristematic tissue, the cambium, which is located just under the bark. It can form bark as well as all the necessary structures that make up the interior body of the plant.

When you take a cutting and place it in the correct environment, it responds by developing new growth. This is particularly so at the nodes, where meristematic tissue is supplied with the auxins that stimulate new root or shoot growth. This can happen at other sites, too, but not every piece of every plant is equally capable of producing a new plant. You'll have the best results by taking the appropriate type of cutting from each plant you are trying to propagate, using rooting hormones that simulate the natural plant auxins that initiate root growth, and placing the cuttings in the correct environments while they are developing.

Lonicera cuttings are best taken from new growth in spring or semiripe growth in summer.

Greenwood cuttings

Greenwood cuttings are slightly older than softwood and slightly younger than semiripe. Like softwood cuttings, they are still in a quick-growing, vegetative phase, so they root very quickly. They are easier to handle, however, because their wood is beginning to become mature, and they don't wilt quite so easily.

Many plants that root well from softwood cuttings do the same from greenwood cuttings. That means if you miss taking a cutting at just the right time to root it from softwood, you can root it as a greenwood cutting. Similarly, if you see a recommendation to root a plant from a softwood cutting, don't hesitate to try it from a greenwood cutting. It won't always work, but you'll have enough successes to make the trial well worth the effort.

Greenwood cuttings are ready at any time from early to midsummer, depending on the plant and your climate. You'll want to take these cuttings at the spot where the current season's growth begins – look for the difference in maturity between last year's bark and this year's growth. Because greenwood cuttings have begun to develop bark, the difference in colouring will be apparent but not startling. If the plant grows fast, you'll be taking as much as 30cm of the branch – but don't let that alarm you. For one thing, nodes below this area will immediately start growing to produce a replacement branch; for another, you'll cut off the top of the new growth (as shown at right) when you prepare it for rooting.

leaves

stem

node

What can go wrong

Fungal diseases: Prevent problems by introducing fresh air into your propagation enclosure several times a day. If problems persist, set up a small computer fan inside the enclosure to keep the humid air moving.

Cutting doesn't root: Environmental conditions must be appropriate. Maintain warm temperatures, high humidity and filtered light in the propagating enclosure.

Checklist

Season: Early to midsummer, depending on species

Tools: Garden knife or sharp bypass pruners, pencil or narrow dibble

Equipment: 10- to 15cm-deep pot or flat, bottomless plastic bottle that fits over the pot or a polyethylene-covered frame; basin for bottom watering

Supplies: Collecting basin or plastic bag and wet paper towels or cloth, plant labels, rooting hormone

Medium: 1 part sand and 1 part vermiculite, or 1 part perlite and 1 part vermiculite

Temperature: Both air and soil should be about 21° to 27°C.

Humidity: Keep humidity high enough to prevent the leaves from wilting, but not so high that they are bathed in moisture. For most cuttings, 70 to 80 per cent humidity is adequate.

Light: Take cuttings in early morning, before the sun is high. Root cuttings in filtered light.

Keep humidity level at 70 to 80 percent so that leaves do not wilt.

1 CUT

When the bottom of the stem has begun to mature, you can use it as a greenwood cutting. Choose shoot terminals and cut at a slant about 1cm below a node. If you are collecting more than one, keep them shaded so they don't wilt.

2 TRIM

To minimize water loss, remove the top, sappy growth of the cutting. Leave at least three nodes. If the leaves are large, cut them in half to reduce water loss.

3 USE HORMONES

Remove the bottom leaf, and wound the bottom of the cutting. Pour a little rooting hormone into a separate container to avoid contaminating your supply, then cover the wound and nodes of the stem with a thin layer. Be certain to cover the nodes and wound thoroughly. Shake off the excess and throw out any unused hormone that you removed from the original bottle.

4 DIBBLE

To avoid wiping the rooting hormone off the cutting, make a hole about 5 to 8cm deep in the medium in which you can insert the cutting. After inserting, push the medium up against the cutting. Set the pot in a basin of water to bottom-water.

5 RETAIN HUMIDITY

Put the pot in a location with indirect light, and arrange the plastic bottle "cover" so it does not touch the leaves. If this is impossible, make a frame of chopsticks and polyethylene to cover the cuttings. If using a bottle, leave the top open to vent excess heat and humidity. If using polyethylene, monitor the interior, and open one side of the enclosure if humidity becomes too high.

6 TUG

After 6 to 8 weeks, most cuttings will have grown roots. Gently tug the stem to see if there is any resistance. If so, carefully remove the cuttings from the pot. If roots are 1 to 2cm long, pot up the cuttings in individual plant pots in a compost-based potting soil, and sink the pots in a garden bed to protect them over the winter months. Mulch the young plants well in the autumn.

Softwood cuttings

Determining what is and what is not "softwood" is much simpler than you may imagine. The branch itself tells you: in early to midsummer, grasp the end and bend it back on itself, about 15cm from the tip, at a 90-degree angle. If it breaks with a sharp, snapping sound, it is at the correct stage to be a softwood cutting. Semi-ripe cuttings won't bend or snap so easily, hardwood cuttings won't bend at all, and branches that are too young to cut won't break.

Softwood cuttings root more easily and quickly than either semiripe or hardwood cuttings, so they are recommended for many plants, particularly hybrids that won't come true from seed. If you follow the guidelines on page 74 and give them an appropriate environment while they're rooting, you can expect much success.

Commercial growers use propagation chambers to root softwood cuttings because they provide the correct environmental conditions. They are fitted with a heating mat to warm the rooting medium, an intermittent mist system to increase the humidity, and walls and a ceiling that allow just enough light to enter so the cuttings can photosynthesize and grow new roots, but not enough to scorch the leaves. If you propagate hundreds of cuttings a year, these units quickly pay for themselves. Most home gardeners, however, are better off improvising to create a similar environment.

If you are rooting only a few cuttings at a time, cut the bottom from a plastic bottle and use it as shown in step 5, opposite. For greater numbers of cuttings, erect a wooden or wire frame, cover it with polyethylene, and set it over the cuttings—making sure they don't touch the plastic. Check the environment frequently. If it is too dry, mist the air with a hand mister several times in the morning and early afternoon. If it is too moist, remove the cover to let the leaves dry and then

APPROPRIATE PLANTS

COMMON	BOTANICAL
Japanese Maple	*Acer palmatum cvs.*
Serviceberries	*Amelanchier* spp.
Japanese Laurel	*Aucuba japonica*
Barberries	*Berberis* spp.
Butterfly Bush	*Buddleja davidii*
Boxwoods	*Buxus* spp.
Beautyberries	*Callicarpa* spp.
Sweetshrub	*Calycanthus floridus*
Blue Mist Shrub	*Caryopteris x clandonensis*
Flowering Quince	*Chaenomeles speciosa*
Summersweet	*Clethra alnifolia*
Redtwig Dogwood	*Cornus alba, C. sericea*
Winter Hazels	*Corylopsis* spp.
Smoke Trees	*Cotinus* spp.
Daphnes	*Daphne* spp.
Slender Deutzia	*Deutzia gracilis*
Burning Bush	*Euonymus alatus*
Wintercreeper	*Euonymus fortunei*
Forsythias	*Forsythia* spp.
Large Fothergilla	*Fothergilla major*
Witch Hazels	*Hamamelis* spp.
Ivies	*Hedera* spp.
Hydrangeas	*Hydrangea* spp.
Virginia Sweetspire	*Itea virginica*
Kerria	*Kerria japonica*
Beauty Bush	*Kolkwitzia amabilis*
Crape Myrtle	*Lagerstroemia indica*
Privet	*Ligustrum japonicum*
Honeysuckles	*Lonicera* spp.
Magnolias	*Magnolia* spp.
Mock Orange	*Philadelphus coronarius*
Chinese Photinia	*Photinia davidiana*
Potentillas	*Potentilla* spp.
Azaleas	*Rhododendron* spp.
Rugosa Rose	*Rosa rugosa*
Willows	*Salix* spp.
Elders	*Sambucus* spp.
Spireas	*Spiraea* spp.
Stewartia	*Stewartia pseudocamellia*
Lilacs	*Syringa* spp.
Viburnum	*Viburnum x burkwoodii, V. carlesii*
Weigelas	*Weigela* spp.

cut vent holes in the top of the enclosure. Check and adjust the temperature, too.

For tender, tropical perennials used as houseplants, the cuttings you'll root aren't technically softwood cuttings because they can be younger and their stems never produce hardwood. For these take tip cuttings, about 10 to 15cm long. Let the stems of very succulent species such as geraniums (*Pelargonium* spp. & cvs.) dry and form a callus, and then root them as described in the steps opposite.

leaves

node

stem

Checklist

Season: Early summer to midsummer, depending on species

Tools: Garden knife or sharp bypass pruners, pencil or narrow dibber

Equipment: 10- to 15cm-deep tray or pot; bottomless plastic bottle that fits over the cuttings, or a polyethylene-covered frame; basin for bottom watering; collecting basin or plastic bag

Supplies: Wet paper towels or cloth

Medium: 1 part sand and 1 part vermiculite, or 1 part perlite and 1 part vermiculite

Light: Take cuttings in early morning, before the sun is high. Root cuttings in filtered light.

Humidity: Keep humidity levels high enough to prevent the leaves wilting, but not so high that the leaves are bathed in moisture. For most cuttings, 70 to 80 per cent humidity is adequate.

Temperature: Both air and soil should be about 21° to 27°C.

What can go wrong

Wilting: Without roots, the cuttings can lose too much moisture from their stems and leaves. Always remove about two-thirds of the foliage from the stem; if the leaves are quite large, cut at least one of the remaining leaves in half.

Rotting: Excessive humidity can lead to fungal diseases. Vent the enclosure at least twice a day, and if humidity levels are excessive, remove the cover overnight. A second cause can be a poorly draining medium. If rotting occurs, add more perlite to the mix, and try again with new cuttings.

1 CUT
Once a stem has stopped elongating – but before it develops hard wood – you can use it as a softwood cutting. Choose shoots that grow from a leader of the plant, and cut at a slant about 1cm below a node. Nestle them in wet paper towels to keep them moist.

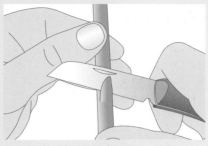

2 SHAVE
Pinch off the bottom leaves, and shave a thin strip of the bark from the bottom of the cutting. These wounds will form calluses that will stimulate cells to produce root tissues.

3 COAT
Coat the bottom of the cutting with a rooting compound as directed on the package, making certain to include the nodes where you've pinched off leaves. Always pour the compound into a separate container and discard it after use to avoid contaminating the material in the jar. Shake off the excess compound before inserting the cutting in the rooting medium.

4 DIBBLE
To avoid wiping off the rooting hormone, make a hole in the medium into which you can insert the cutting, about 5 to 8cm deep. Push the medium up against the cutting. Set the pot in a basin of water to bottom-water.

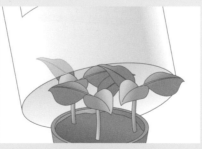

5 COVER
Put the pot in a location with indirect light, and arrange the plastic bottle "cover" so it does not touch the leaves. If this is impossible, make a frame of chopsticks and polyethylene to cover the cuttings. If using a bottle, leave the top open to vent excess heat and humidity. If using polyethylene, monitor the interior and open one side of the enclosure if humidity becomes too high.

6 TUG
After 6 to 8 weeks, most cuttings will have grown roots. Give a gentle tug to the stem to see if there is any resistance. If so, carefully remove the cuttings from the pot. If roots are 1 to 2cm long, pot up the cuttings in individual plant pots in a compost-based potting soil, and sink the pots in a garden bed to protect them over the winter months. Mulch the young plants well in the autumn.

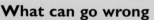

Semiripe cuttings

Conifers, broad-leaved evergreens and many shrubs are frequently propagated by rooting semiripe cuttings. In the case of conifers and broad-leaved evergreens, this is almost always the preferred method of vegetative propagation. It's highly effective with most shrubs, too, so it's worth a try when you can't find information about propagating a shrub.

Semiripe cuttings are older than greenwood cuttings but younger than hardwood cuttings. Again, they are taken from the current season's growth. To be certain that you are getting them at the correct stage of growth, start watching the plants in early summer – some plants in some climates produce semiripe cuttings as early as June, while others do so in July or even early August.

These cuttings are characterized by soft growth at the stem tip and much harder growth about 15 to 20cm lower. Cut below the bottom growth, just below a leaf node – you'll want some of this hardening wood – but above the previous year's growth. If possible, choose stems without developing buds or blooms; you want the stem to put maximum energy into developing a root system.

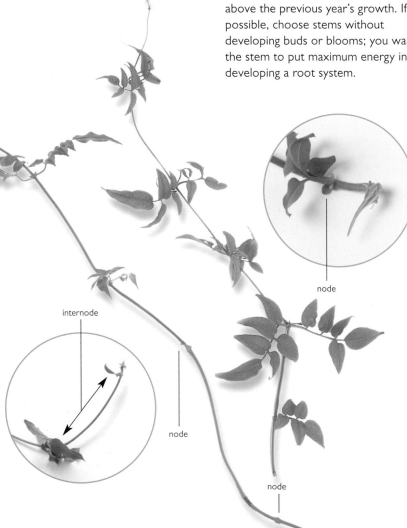

internode

node

node

node

What can go wrong

Fungal diseases: Fungi cause problems with all types of cuttings. Be certain that your rooting medium is fresh. Don't let humid air stagnate around the cuttings.

Cuttings don't root: Cuttings won't root in a medium that is too dry. Check frequently to make certain that it is consistently moist.

Checklist

Season: Late spring to midsummer, depending on species

Tools: Garden knife or sharp bypass pruners, pencil or narrow dibble

Equipment: 10- to 15cm-deep pot or tray, bottomless plastic bottle that fits over the pot or a polyethylene-covered frame; propagation chamber or soil-heating mat for some species; basin for bottom watering

Supplies: Collecting basin or plastic bag and wet paper towels or cloth, plant labels, rooting hormone

Medium: 1 part sand and 1 part vermiculite, or 1 part perlite and 1 part vermiculite

Temperature: Both air and soil should be about 21° to 27°C.

Humidity: Keep humidity levels high enough to prevent the leaves wilting, but not so high that they are bathed in moisture. For most cuttings, 70 to 80 per cent humidity is adequate.

Light: Take cuttings in early morning, before the sun is high. Root cuttings in filtered light.

Destroy cuttings that have fungal disease.

1 CUT
Cut semiripe cuttings from a plant when 15- to 20cm long shoots have begun to "ripen", or become woody, at the base.

2 TRIM
Trim off all sideshoots and the leaves from the bottom 4cm of the stem. If top growth is soft and sappy, trim it, too.

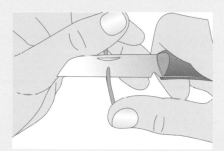

3 WOUND
Wound the cutting by removing a 2cm-long, slender strip of bark from the base of the cutting. The resulting callus that forms over the wound will stimulate the formation of roots.

4 TREAT
Pour a little rooting hormone in a separate container to avoid contaminating your supply, and then cover the wound and nodes of the stem with a thin layer. Shake off the excess and throw out any unused hormone that you removed from the original bottle.

5 DIBBLE
Dibble holes in your planting media and place the cuttings into the holes. Firm the medium around the cuttings and water well. Place the plastic bottle cover over the pot (not shown).

6 TEST
After about 2 months, gently tug on a bottom leaf to see if the plant has rooted. If so, unpot the plant. It will be ready to transplant into larger pots when the roots are about 2cm long. Keep the pots cool over the winter; in spring, plant them in a garden nursery bed.

Hardwood cuttings

After autumn leaves have dropped and woody plants have gone dormant is the time to take hardwood cuttings. For many woody plants, this is the easiest way to ensure propagation success, partly because they are so impervious to damage at this stage and partly because the callus that forms on the end of the stem over winter stimulates root production.

There are three types of hardwood cuttings: straight, heel and mallet. Most deciduous shrubs and trees – and many evergreen broad-leaved plants – root well from straight hardwood cuttings, but occasionally you'll run into directions that advise working with a heel or mallet cutting when propagating a deciduous plant. Directions for these cuttings are on pages 84 and 85.

When taking a straight hardwood cutting, cut just below a node of the past season's growth – you'll still be able to distinguish this wood. Depending on the plant, you may get more than one cutting from one branch; each cutting should include about five nodes. In contrast to other stem cuttings, make the bottom cut straight across the wood and the top cut slanted. To minimize the threat of fungal infection, cut off the softer wood at the tip; you want only fully hardened wood.

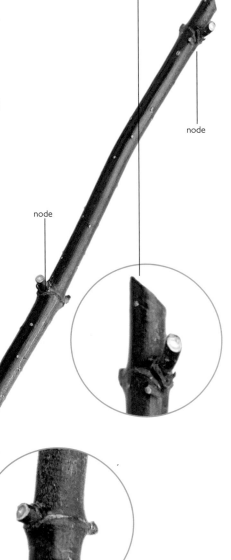

slanted cut

node

node

node

nodes

straight cut

APPROPRIATE PLANTS

COMMON	BOTANICAL
Abelias	*Abelia* spp.
Kiwifruit	*Actinidia* spp.
Butterfly Bushes	*Buddleja* spp.
Camellias	*Camellia* spp.
Trumpet Vines	*Campsis* spp.
Dogwoods	*Cornus* spp.
Quince	*Cydonia oblonga*
Deutzias	*Deutzia* spp.
Escallonias	*Escallonia* spp.
Figs, ornamental and edible	*Ficus* spp.
Forsythias	*Forsythia* spp.
Franklinia	*Franklinia alatamaha*
Silk Tassel Bushes	*Garrya* spp.
Ivies	*Hedera* spp.
Rose of Sharon	*Hibiscus syriacus*
California Spirea, Cream Bush	*Holodiscus discolor*
Hydrangeas	*Hydrangea* spp.
Jasmines	*Jasminum* spp.
Junipers	*Juniperus* spp.
Japanese Kerria	*Kerria japonica*
Himalayan Honeysuckle	*Leycesteria formosa*
Privets	*Ligustrum* spp.
Honeysuckles	*Lonicera* spp.
Virginia Creepers, Boston Ivies	*Parthenocissus* spp.
Mock Oranges	*Philadelphus* spp.
Cherries, Peaches	*Prunus* spp.
Rhododendrons	*Rhododendron* spp.
Roses	*Rosa* spp.
Willows	*Salix* spp.
Lavender Cottons	*Santolina* spp.
Stephanandras	*Stephanandra* spp.
Viburnums	*Viburnum* spp.
Grapes	*Vitis* spp.
Weigelas	*Weigela* spp.

What can go wrong

Cuttings rot: Fungal pathogens can attack hardwood cuttings. Be certain to use fresh, clean starting media; if starting more than one cutting, remove any that look diseased to avoid infecting others.

Cuttings don't root: Hardwood cuttings usually require rooting hormone. Check to be certain that you are using a compound meant for hardwoods at the correct concentration; too much, and rooting will be inhibited. You may also need to wound the stem by cutting off a sliver of bark about 2cm long at the bottom end. This gives a greater area of callus and thus more opportunity for roots to form.

Very slender hardwood cuttings don't usually root well. Try to use cuttings about as thick as a pencil.

Checklist

Season: Late autumn or early winter, after the plant is dormant

Tools: Bypass pruning shears, sharp garden knife

Equipment: Pot or 10cm-deep box or tray, cold frame

Supplies: Hardwood rooting compound, plant labels

Medium: Sharp sand or a mixture of sharp sand and perlite

Temperature: Almost all species callus well at 4°C. In spring, temperature preferences vary according to the climatic preferences of the particular plant.

Light: Diffused or semi-shade

1 CUT
Ideally, hardwood cuttings should be about as thick as a pencil. After taking the cutting, remove any softwood at the top, again cutting above a node. Make your bottom cut straight across and your top cut at a slant. Don't let the cuttings dry – wrap them in damp paper towels until you can plant them.

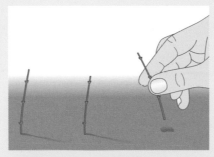

3 PLANT
The following spring, plant the cuttings in a sand-filled trench in a nursery bed in the garden. Firm the sand well so there are no air pockets around the stems. Roots will grow from along the stems and at the cutting ends.

5 DISCARD
By midsummer, you'll be able to tell which cuttings put on enough root growth to support new growth and which didn't. Remove any cuttings that did not sustain their leaf growth; this ensures that any secondary pathogens attracted to the dying cutting don't have an opportunity to build to damaging population levels.

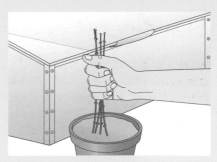

2 BURY
Gather the cuttings into bundles, and treat the stem bottoms with a rooting compound. Bury the bottom two-thirds of their length in the rooting medium. If you live in a relatively warm climate (Zones 8 to 10), you can place the cuttings outside without cover for the winter months. In moderately cool areas, place them in cold frames that you can keep insulated and out of the wind during the winter. In Zones 5A and northward, it's best to lay cuttings in a sand-filled box, water the sand, and then bury the box about 30cm under the soil.

4 MONITOR
Keep the sand consistently moist, and monitor the cuttings on a regular basis. Some will develop leaves within a few weeks. Provide afternoon shade if the cuttings are not in an area with filtered light.

6 TRANSPLANT
Depending on the speed with which the cutting grew and the severity of your climate, you can transplant the new plant into its permanent position in the autumn or wait until the following spring. If you live in a cold-winter area, remember to mulch the plant well after the ground freezes.

Conifer cuttings

Although conifers grow well from seeds, many gardeners prefer to propagate them from cuttings. Seeds are the more reliable propagation method, but cuttings have two notable advantages: they produce larger plants more quickly than seeds do and the new plants have the same characteristics as the parent plant. Any differences in the mature plants will be due entirely to environmental differences – the cutting-grown plants are clones of the parent, and their genetic code is identical.

Cuttings from conifers are called hardwood because they are usually taken in late autumn to early winter, but they are treated more like other types of cuttings because they retain their leaves and continue to photosynthesize, respire and transpire – all without roots – just as softwood, greenwood and semi-ripe cuttings do.

Take conifer cuttings from side shoots growing in the middle to the lower sections of the shrub or tree; cuttings from the top portions do not generally root as well as those from lower growth.

Many juniper species are resistant to rooting. To get around this problem, take a mallet cutting – one that includes about 1 to 2½cm of the branch from which the stem grows when you take the cutting. This wood is 2 years old and, oddly enough, roots more easily than straight cuttings of junipers. Arborvitae (*Thuja*) can also present problems, but in their case, take a heel cutting – a sliver of the bark and some of the interior of the wood from which the cutting emerges. Heel cuttings are also useful with other conifers.

Checklist

Season: Late autumn or early winter

Tools: Bypass pruning shears, sharp garden knife

Equipment: Pot or tray at least 10cm deep, insulated cold frame or propagating chamber with bottom heat and intermittent misting or homemade frame with soil-heating mat and misting unit on a timer

Supplies: Hardwood rooting compound, shade cloth, plant labels

Medium: Half peat moss or leaf mould and half sand

Temperature: Keep a propagating chamber at about 20°C. If rooting cuttings in a cold frame in the garden, maintain day temperatures of at least 16°C and night temperatures of 2° to 4°C.

Light: Bright in northern regions; filtered in bright, southern areas

mallet cutting

needles

heel cutting

stem

1 CUT

You'll have the best chance of success if you select a horizontal branch that isn't more than 5 years old and that's at least 10 to 13cm long. Strip the bottom leaves from the cutting.

2 DIP

Remember to use hardwood rooting compound. Dip your stems in some that's been poured out of the original container, and shake off the excess.

3 DIBBLE

Place the cuttings in prepared holes in the medium, and firm the medium around them to make certain there are no air pockets.

4 LET ROOT

Place the pot in a propagating chamber with bottom heat or in a cold frame in a protected area outdoors. If outdoors, erect a shade cloth or piece of hessian over the cold frame to provide filtered shade.

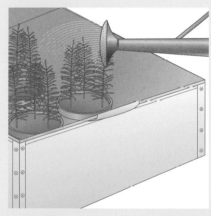

5 MONITOR

Monitor the cuttings all winter to be certain that the medium stays moist and temperatures remain within the preferred range. Cuttings that were kept inside all winter will develop roots in about a month; those outside may wait until spring.

6 GROW ON

Once inside cuttings have roots, transplant them into a humus-rich soil mix and larger pots. Once the frost-free date has passed, move the pots outside. Monitor the outside cuttings to determine when their root systems are large enough to handle transplanting to larger quarters and a humus-rich soil mix. In autumn, sink the pots in the garden. By the following spring, the new plants will be ready to transplant into their permanent positions.

What can go wrong

Cuttings rot: As with all cuttings, fungi are the most likely source of problems. Keep air circulation high in a propagating chamber, and no matter what the location, don't water so much that the medium becomes saturated.

Fungal disease can take a toll on cuttings.

Cane cuttings

You probably won't use this technique often (unless you live in a tropical area where these plants grow naturally). Nonetheless, it's nice to know how to propagate them. If nothing else, you can develop what feels like an indoor jungle to hold the winter blues at bay, or grow pots and pots of these exotic beauties and donate them to charity plant sales.

Plants with canes are so easy to propagate that your only problem is likely to be one of oversupply. To root them, simply cut a cane into 5- to 10cm lengths, and lay these pieces horizontally on the rooting medium. Once the nodes of these plants are in touch with moist soil mix, they are stimulated to produce roots.

This is an ideal technique to share with young children. However, remember that dumb cane gets its common name for a reason – the sap that flows out when the cane is cut can stimulate an allergic rash, numb the mouth, and be lethal if eaten. Wear latex gloves when you work with dumb cane, and choose another plant if you want to introduce your children to this easy propagation technique.

APPROPRIATE PLANTS

COMMON	BOTANICAL
Chinese Evergreens, Ribbon Evergreens	*Aglaonema* spp.
Ti Plant	*Cordyline fruticosa*
Dracaena	*Cordyline indivisa*
Dumb Canes	*Dieffenbachia* cvs.

Checklist

Season: Any season

Tools: Sharp garden knife, dibble

Equipment: Tray; 5- to 8cm-deep or wide pot; soil-heating mat; plastic-covered frame or plastic bottle

Supplies: Plant labels

Medium: Fast-draining soil-less medium such as 1 part perlite and 1 part peat moss for rooting; soil mix for growing on

Temperature: Minimum of 21°C

Humidity: Relatively high; at least 40 per cent

Light: Bright in northern regions; filtered in bright, southern areas

nodes

What can go wrong

Canes rot: Make certain that the medium is moist enough to supply developing roots with water but not so wet that it promotes the growth of fungal diseases.

Canes don't root: Keep both air and soil temperatures high while rooting canes. These are subtropical or tropical plants and prefer temperatures of at least 26°C.

In a medium kept this moist, the cane is almost sure to rot.

Cordyline fruticosa can be propagated using cane cuttings or tip cuttings. When taking cane cuttings, ensure you have two joints along the stem.

1 CUT

Cut off some of the top growth or, if you want to thin the plant, the entire cane.

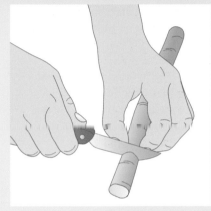

2 TRIM

Canes as short as 5cm long root well, as long as they contain at least one joint. But to be on the safe side, take cuttings with two joints until you are certain that your environment is appropriate.

3 BURY

Place the canes horizontally on the tray, burying them no more than half their width.

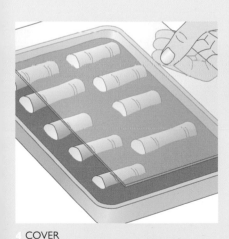

4 COVER

Humidity levels must remain high while the cane is rooting.

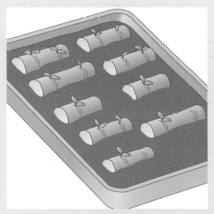

5 MONITOR

Turn off the soil-heating mat, and remove the plastic cover as soon as it becomes apparent that shoots are forming. Excess humidity at this stage could promote fungus diseases.

6 PLANT

Once the plant has roots and a strong shoot, pot it up in a fast-draining soil mix in an individual pot. Keep transplanting as the plant outgrows successive pots.

Leaf petiole cuttings

The petioles of some leaves are capable of growing whole new plants. But as you'll see by looking at the list of appropriate plants, right, this ability is restricted to tropical and subtropical plants with fleshy leaves.

Growing new plants from petioles is a simple propagation technique. As long as your environment is suitable, it's almost guaranteed. If you hanker for the garden all winter long or want to introduce your children to the joys of propagating plants, try working with leaf petiole cuttings.

A sharp knife is mandatory for success. You'll be cutting the petiole at an angle, and a blunt knife could injure it badly enough to prevent it from transporting water or nutrients. If you're extremely dexterous, you can use a razor blade. Otherwise, a long-handled utility knife (available from any artist's supply store) does the trick.

APPROPRIATE PLANTS

COMMON	BOTANICAL
Jade Plant	*Crassula ovata*
Flame Violets	*Episcia* spp.
Hoyas	*Hoya* spp.
Peperomias	*Peperomia* spp.
African Violets	*Saintpaulia* cvs.
Sedums	*Sedum* spp.

Checklist

Season: Any season

Tools: Sharp scissors, razor blade, or long-handled utility knife; dibble

Equipment: 10cm pot, plastic bottle for cover, 6cm pot

Supplies: Plant labels

Medium: 1 part perlite and 1 part vermiculite for rooting; soil mix for growing on

Temperature: Minimum of 24°C

Humidity: Relatively high; at least 50 per cent

Light: Diffused

petiole

angled cut

leaf

Insert the petiole into the rooting medium.

petiole

What can go wrong

Cutting rots Like all cuttings, these are susceptible to various fungal diseases. Avoid problems by keeping the rooting medium moist but not soggy. If you are covering the plants to keep relative humidity levels high, remember to introduce fresh air several times a day. Don't allow moisture to condense on the covering, either – it will drip onto the leaf and promote problems.

This medium is so wet that the petiole is likely to rot.

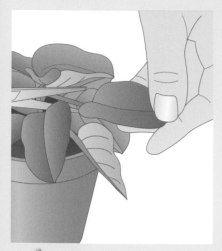

1 CUT

Use sharp scissors to cut a young but full-size leaf, including its petiole, from the parent plant.

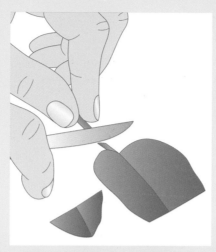

2 TRIM

Remove some of the top of the leaf, then cut the petiole to a length of about 1 cm. Next, cut the bottom of the petiole at an angle of about 45 degrees to expose the interior to the rooting medium.

3 DIBBLE

Make a hole for the petiole so you don't damage it when you insert it into the medium. Gently firm the medium around it.

4 COVER

Place the cutting in a moderately warm spot with diffused light. If indoor humidity is low, cover the pot while the cutting is rooting. Remember to lift the bottle to allow fresh air into the enclosure several times a day.

5 SEPARATE

In about 4 months, the plantlets that have grown from the petiole are large enough to be separated. Gently tease them apart, remembering to touch only the bottom leaves of each one.

6 TRANSPLANT

Transplant your plantlets into your normal soil mix when they seem large enough to be handled without injury.

Leaf vein cuttings

The first time you propagate plants from the veins of a leaf, you're likely to be amazed – the sight of a group of baby plantlets growing out of a leaf is so surprising! This technique is easy when the environment is suitable, but if you have difficulty maintaining high indoor-humidity levels in winter or keep your home much cooler than 21°C, you'll be better off waiting until warmer weather.

APPROPRIATE PLANTS

COMMON	BOTANICAL
Eyelash Begonia	*Begonia bowerae*
Reiger Begonia	*Begonia* × *hiemalis*
Iron Cross Begonia	*Begonia masoniana*
Rex Begonia	*Begonia rex-cultorum*
Venus Flytrap	*Dionaea muscipula*
Sundews	*Drosera* spp.
Florists' Gloxinias	*Sinningia* spp.
Smithianthas	*Smithiantha* spp.

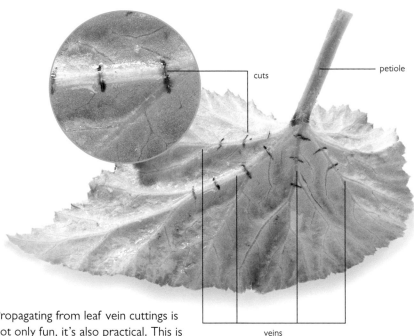

cuts

petiole

veins

Checklist

Season: Any season, but spring and summer are most reliable

Tools: Sharp scissors, razor blade, or long-handled utility knife; dibble

Equipment: Small paintbrush, 15- to 20cm pot, hairpins, plastic bottle or plastic-covered wire frame for cover, 6cm pots

Supplies: Rooting hormone, plant labels

Medium: Sterile soil-less mix for rooting, humus-rich soil mix for growing on

Temperature: Minimum of 24°C

Humidity: Relatively high; at least 55 per cent

Light: Diffused

Propagating from leaf vein cuttings is not only fun, it's also practical. This is particularly so in the case of rex begonias because they vary so much from one to another. If you happen to have a plant that you love, propagating it from a leaf assures you of being able to grow many more of the same sort.

Sundews and Venus flytraps – which can also be propagated with this technique – are carnivorous plants that secrete a digestive enzyme from their leaves. When you touch them to cut through the veins, you're likely to come into contact with it. Such a small concentration shouldn't harm you, but if you have sensitive skin, you might want to wear latex gloves while working with these plants.

What can go wrong

Leaf rots: Keeping the humidity high enough to promote the growth of new plants but not so high that they succumb to a fungal disease can be a challenge. If the humidity is under 55 per cent, use plastic sheeting or an open plastic bottle to retain humidity. Keep a constant check on it, and open the enclosure several times a day to let in fresh air.

Nothing grows: The medium must be kept moist but not soggy, and both air and soil temperatures must be 24° to 29°C to stimulate growth. If in any doubt about the temperature, place the pot on a soil-heating mat, and enclose the pot and mat to keep interior temperatures high.

If you cut entirely through the vein, the end of the leaf will die before new plants can form.

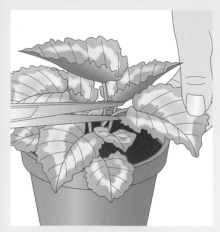

1 CUT
Cut a healthy, mature leaf from the parent plant. This leaf should not be one of the oldest, but it should be full-size.

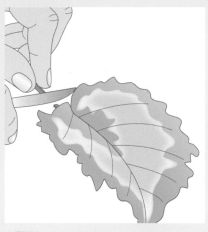

2 TRIM
Cut off the petiole flush with the bottom of the leaf.

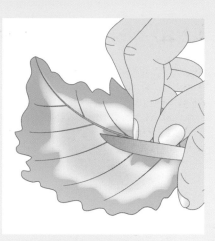

3 CUT AGAIN
Turn the leaf over on a sterilized cutting board, and cut through several of the larger veins. Make the cuts about 2cm apart, and do your best not to cut through the entire leaf. After cutting, use a small paintbrush to dust the cut surfaces with rooting hormone.

4 PEG
Turn the leaf right-side up, and set it in the pot so the cut veins are in contact with the dampened soil-less medium. Use U-shaped hairpins, not hair grips, to hold the leaf in place – place them so they straddle the veins.

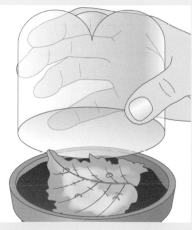

5 COVER
Use a cover to maintain high humidity levels if necessary; monitor temperatures and humidity on a consistent basis, adjusting when needed.

6 TRANSPLANT
As soon as the new plants have their second set of leaves, they can be transplanted into individual pots. If they don't pull apart easily, cut the parent leaf into sections, each containing a new plant.

Begonia plants are easily propagated from leaf vein cuttings.

Upright leaf cuttings

The principle of propagating new plants from upright leaf cuttings is the same as that of propagating them from a leaf pressed against the surface of a moist soil-less medium – it's appropriate for plants with the ability to produce both roots and shoots from a leaf vein. However, some plants root more easily from leaf sections that are held upright by the rooting medium, so the methods aren't entirely interchangeable.

In the correct environment, this propagation method is just about guaranteed to produce new plantlets. In some cases, such as with strepto-carpus cultivars, you may be stunned by your success – most people get enough plants from a single pot of leaf sections to supply the neighbourhood. Other plants are more restrained; you're likely to get only one or two jade plants (*Crassula argentea*) from a

leaf section, for example.

Choose your parent leaves well. It's generally best to use leaves that are growing vigorously and have just reached their mature size. Young leaves are more susceptible to fungal infections, and old leaves are less likely to produce new plants. Always check to make sure that the leaf is attractive, too. Remember, the plants that you'll be getting will look like the parent leaf.

APPROPRIATE PLANTS

COMMON	BOTANICAL
Eyelash Begonia	*Begonia bowerae*
Reiger Begonia	*Begonia x hiemalis*
Iron Cross Begonia	*Begonia masoniana*
Rex Begonia	*Begonia rex-cultorum*
Jade Plant	*Crassula argentea*
Kalanchoes	*Kalanchoe* spp.
Prayer Plants	*Maranta leuconeura*
African Violet	*Saintpaulia* hybrids
Streptocarpi	*Streptocarpus* spp.

Checklist

Season: Any season, but spring and summer are most reliable

Tools: Sharp scissors, razor blade, or long-handled utility knife; dibble

Equipment: Small paintbrush, 5cm-deep seed tray or 15cm pot, plastic bottle or plastic-covered wire frame for cover, 6cm pots

Supplies: Rooting hormone, plant labels

Medium: Sterile soil-less mix for rooting, humus-rich soil mix for growing on

Temperature: Minimum of 24°C

Humidity: Relatively high; at least 55 per cent

Light: Diffused

Place the leaf sections in the medium so that only the bottom of the leaf is buried.

leaf tops

petiole

If too much of the leaf section is buried, it runs more risk of rotting. Keep about 80 percent of the leaf section above the medium

1 CUT
Cut off the petiole from the leaf you've chosen to propagate.

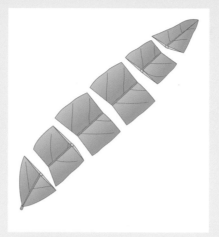

2 SECTION
Cut the leaf crosswise into sections about 5cm wide. To avoid turning them upside down, leave them on the cutting board until you place them in the medium. If desired, brush the bottoms with a small amount of rooting hormone, but this usually isn't needed.

3 PREPARE MEDIUM
Create furrows about 5 to 7½cm apart in the moistened rooting medium.

4 PLANT
Place the leaf section so that its "bottom" is inserted into the medium and its "top" is upright. Firm the medium around the leaf section to keep it upright.

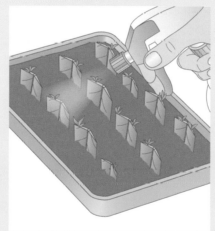

5 MONITOR
Remove the plastic cover after the new plantlets become visible. Water with a hand mister to keep the medium moist but not saturated as the plants grow.

6 TRANSPLANT
After the plantlets have two sets of leaves, transplant them to individual pots.

What can go wrong

No plantlets form: In cool conditions, plantlets may not form. Set the pot or tray on a soil-heating mat, and enclose it in a plastic-covered wire frame if air and soil temperatures are below 24°C.

Cuttings rot: As always, fungal diseases are the enemies of cuttings. Make certain that the rooting medium is sterile to begin with, and monitor soil moisture and relative humidity levels.

Monocot leaf cuttings

Monocots are plants such as onions, lilies, maize and grasses that have only one cotyledon, or embryonic leaf, in their seeds rather than the two that dicots have. Another distinction is the arrangement of the leaf veins: in dicots, the veins branch; in monocots, the veins are parallel to each other and run vertically on the leaf.

Plants such as lilies and hyacinths are much easier to propagate with bulbs and bulblets than with leaf cuttings, so it's sensible to stick with bulbs. However, in the case of the plants listed above right, leaf cuttings are a practical propagation method.

Mother-in-law's tongue, or snake plant (*Sanseviera trifasciata*), produces seeds in a warm, sunny environment and offsets even in a pot on the windowsill. However, seeds are unreliable in northern areas, and offsets are slow to develop, so leaf cuttings tend to be the preferred method.

Contrary to what you might expect, the plants that grow from leaf cuttings do not look identical to the parent leaf. Instead, they tend to be a solid green. If you want a plant with the parent's colouring, you'll have to divide an offset from the parent's rhizome.

Checklist

Season: Any season, but spring and summer are most reliable

Tools: Sharp utility knife, dibble

Equipment: 15- to 20cm pot or seed tray, plastic bottle or plastic-covered wire frame for cover

Supplies: Plant labels

Medium: Clean sand, fast-draining soil mix for growing on

Temperature: Minimum of 24°C

Humidity: Relatively high; at least 55 per cent

Light: Diffused

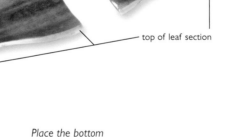

bottom of leaf section

top of leaf section

Place the bottom of the leaf in the medium.

What can go wrong
Leaf section rots: Although the plants that grow well from these leaf cuttings thrive in high-humidity conditions, none of them tolerates soggy soil. Root them in a fast-draining medium, and don't let moisture condense on the ceiling of their enclosure.

Don't bury the leaf section too deeply or upside down. Leave 80 percent of the leaf section above the medium.

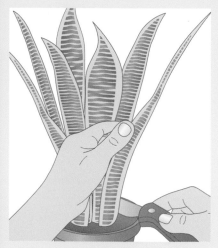

1 CUT

For the sake of the parent plant's appearance, remove an entire leaf when you are taking it for cuttings.

2 SECTION

Cut the leaf into sections about 5 to 10cm long. Don't scramble your pieces – they won't root upside down.

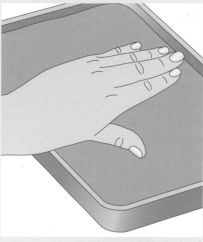

3 PREPARE MEDIUM

A soggy medium kills this plant; use moist sharp sand to root the cuttings.

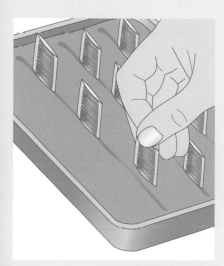

4 PLANT

Place the bottoms of the cuttings in preformed furrows, and firm the medium around them to hold them in place.

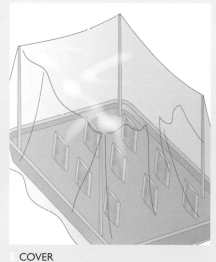

5 COVER

Enclose the flat to retain humidity. If you don't have a frame, use chopsticks to suspend the plastic covering.

6 TRANSPLANT

You'll see the first roots in about a month, but it will take about 2 months for plantlets to get large enough to transplant into their own pots.

Root cuttings

Weeds in your garden have probably demonstrated how easily some plants propagate themselves from a tiny piece of root. The common herbs comfrey and milk thistle, for example, can turn into weeds if you make the mistake of moving pieces of their roots around with a tiller. Not all the plants you'll propagate this way are so invasive, but they all root easily from cuttings of their roots.

Take root cuttings in late autumn or early winter, when the plants are dormant. At this time, the carbo-hydrate levels in the root are high enough to see it through its dormant period. This food supply will sustain it through the rooting period as well, making it much easier for the plant to form roots and shoots in time for the coming season.

Some of the species that grow well from root cuttings produce new shoots before making new roots – the shoots grow the roots that will eventually sustain them. But other plants develop root systems before producing new shoots. This process takes longer to be visible, so be patient while you're waiting for proof that a root cutting has taken.

APPROPRIATE PLANTS

COMMON	BOTANICAL
Bear's Breeches	Acanthus spp.
Bugbanes	Actaea spp. (syn. Cimicifuga spp.)
Windflowers	Anemone spp.
Bergenias	Bergenia spp.
Siberian Bugloss	Brunnera macrophylla
Trumpet Vines	Campsis spp.
Cupid's Dart	Catananche caerulea
Giant Scabious	Cephalaria gigantea
Bleeding Hearts	Dicentra spp.
Gas Plants	Dictamnus spp.
Purple Coneflower	Echinacea purpurea
Globe Thistle	Echinops ritro
Sea Hollies	Eryngium spp.
Figs, Ficuses	Ficus spp.
Blanket Flowers	Gaillardia spp.
Hops	Humulus spp.
Golden Rain Trees	Koelreuteria spp.
Leopard Plants	Ligularia spp.
Sea Lavenders, Statices	Limonium spp.
Osage Orange	Maclura pomifera
Sundrops	Oenothera spp.
Oriental Poppies	Papaver spp.
Empress Tree	Paulownia tomentosa
Phlox	Phlox paniculata
Chinese Lanterns	Physalis spp.
Primroses	Primula spp.
Roses	Rosa spp.
Milk Thistle	Silybum marianum
Stokes Aster	Stokesia laevis
Comfreys	Symphytum spp.
Lilacs	Syringa spp.
Mulleins	Verbascum spp.

root

top of root

A straight cut at the top of the root and a slanted cut at the bottom help to distinguish the ends.

bottom of root

What can go wrong

Root cutting doesn't develop: Most root cuttings take easily, but in cases where one doesn't, suspect that the medium was so moist that disease struck or that its temperature was either too high or, more commonly, too low.

Very slender roots don't have enough stored carbohydrates to be able to produce a new plant.

Checklist

Season: Late autumn, early winter

Tools: Garden trowel or spading fork, bypass pruners, sharp garden knife, dibble

Equipment: 15- to 20cm pot, box filled with moistened peat moss or sand, cold frame

Supplies: Plant labels

Medium: Soil-less potting mix

Temperature: Minimum of 21°C

Humidity: Not important

Light: Not important

1 DIG

With care, you can take root cuttings from plants without removing them from the soil. Dig on one side to expose a few roots that are about as thick as a pencil, and cut one or two of these off, close to the crown.

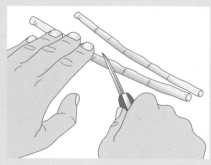

2 CUT

Cut the roots into pieces 5 to 15cm long. Make a straight cut on the top of the root and a slanted cut on the bottom so you can distinguish the ends later.

3 STORE

Bundle the roots, all facing the same direction, and place them in a box filled with moistened peat moss or sand. Store this box in the refrigerator or in another dark area that maintains temperatures of around 4°C for about 3 weeks.

4 PLANT

Plant the root cuttings in a pot with the top ends placed just under the soil surface. Keep the cuttings in a cool but not freezing place, such as a cold frame or basement, until spring.

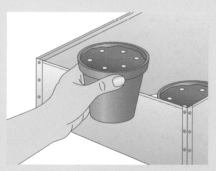

5 MOVE

If you've stored the pots indoors over the winter, move them to a cold frame in early spring so they can begin to experience normal temperature fluctuations of the season.

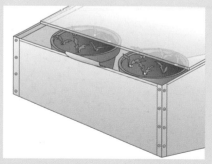

6 TRANSPLANT

Roots and shoots will form within a month or two. Transplant the plantlets to pots containing a nutrient-rich soil mix, and grow on until early autumn, when you can transplant them to permanent positions in the garden.

LAYERING

Layering is one of the easiest ways to propagate plants. The stems of many plants naturally form roots when they are in contact with the soil, and this method takes advantage of that characteristic. With some plants, you have to wound the stem to stimulate rooting, and you often have to bury or surround the stem with soil or a moist rooting medium, but that's all there is to it — the plant does the rest of the work.

The layered area remains attached to the parent plant while it is rooting, so if it has enough water to stimulate rooting, it's unlikely to die from drought afterward. Similarly, because it is growing in the same environment as its parent is, it's unlikely to develop diseases on its own.

When you look in the Plant Directory (page 134), you'll notice that you have a choice of propagation methods for many plants. As you consider these options, remember that "sure and simple" techniques such as layering have many advantages. Propagating with a simple technique may not be as thrilling as successfully carrying out a difficult method, but it's likely to take less time or skill to accomplish and usually produces more certain results. This principle holds even when you have a choice between various layering techniques; if you can choose between simple layering and dropping, take the easy path — simple layering — if you're after a sure outcome.

The directions for the various layering procedures are all somewhat different. However, the commonsense guidelines on the facing page are always applicable.

Top left: Viburnum can be propagated by layering in the spring.
Above: Layering lavender in mid spring ensures the same type of lavender is propagated.

Below: Once roots have formed on a layered stem, it can be cut from the parent.

When layered, young Hedera forms adventitious rootlets along its nodes.

Guidelines for success

• Layer a plant when it is in an active growth phase.

• Loosen the soil where the layered stem will form its new root system; if the soil is clayey or compacted, dig in some fully finished compost and sand so the roots will have an easy time growing into it.

• Strip leaves off the stem to be layered for 8 to 10cm above and, when appropriate, below the area that you put into contact with the soil or rooting medium. This allows the plant to put maximum energy into making new roots.

• Choose stems from the current or previous season's growth.

• Wound the part of the stem that is in contact with the soil or medium by cutting a 5cm slash about a third of the way into the wood. Push a small stone or wooden matchstick into the wound to keep it open when the stem is buried.

• If the stem seems likely to pull out of the soil, set a heavy rock over the buried portion.

• You can insert a slender stake next to the part of a layered stem that is not buried and tie the stem to it – this will encourage the new plant to grow upright.

• Keep the area around the buried stem weeded and moist.

• Leave the plant attached to its parent until it is putting on new, strong growth – this may take 12 to 18 months. After severing the new plant, leave it in place for another season before subjecting it to transplanting.

Raspberry plants often layer themselves – without any help from you!

Tip layering

If you've grown trailing blackberries, you've probably seen natural instances of tip layering. When you don't keep up with the trellising on these plants, a few stems usually arch forward and bend to the ground. If a stem tip sits on the soil surface, it's likely to take root there. Within a year or so, you can sever this tip-layered plant from its parent.

APPROPRIATE PLANTS

COMMON	BOTANICAL
Forsythias	*Forsythia* spp.
Blackberries	*Rubus* spp.
Raspberries	*Rubus* spp.

Tip layering is the easiest way to propagate raspberries and blackberries. Because this technique is so simple, you might think that it would also be the best way to propagate other vining plants. Oddly enough, it doesn't work well with many plants other than those in the *Forsythia* and *Rubus* genera; most vining plants and trailing shrubs grow best from simple or serpentine layering, as described on page 102.

Bramble fruits are susceptible to numerous diseases, including various viruses that sucking insects such as aphids and leaf hoppers transmit. If your plants' yields have been declining, their leaves show odd puckering or yellowing or bushy growth emerges from their stem tips, it's likely that they are hosting a viral infection. Do not propagate them – you'll only increase the population of the disease. Instead, remove your diseased specimens. Buy new, certified virus-free plants, and propagate these.

stem tip

parent stem

A new shoot will grow from this parent stem.

1 DIG
Dig a small hole in the soil where you'll bury the stem tip of the plant you're going to propagate. Remember that you can propagate many stems at once if you need to increase your planting radically.

2 PLACE
Place the tip of the stem in the hole you've dug. If the stem is thorny, wear gloves to protect yourself from scratches.

3 BURY
Bury the tip of the stem at least 2½cm below the soil surface, and tie the parent stem to a stake to hold the tip in place.

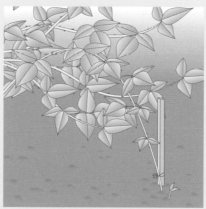

4 MONITOR
You will see a new shoot growing from the parent stem in only a few weeks.

5 ALLOW TO GROW
Cut the parent stem if you wish, but allow the new plant to grow in place for a season.

6 TRANSPLANT
Transplant the new plant the following spring – by then it will have a large root ball and can survive easily on its own.

Checklist
Season: Mid to late summer

Tools: Trowel, spade, bypass pruners

Equipment: Gloves to protect against thorns

What can go wrong
New plant dies: Don't sever the plantlet from its parent cane too early. If you have any doubts about its ability to survive on its own, brush aside the surrounding soil to investigate the root system. If the root system is still small, leave it in place until the following spring.

If the medium remains too wet, the tip layer may rot.

Simple and serpentine layering

It's almost always worth trying to layer a favourite plant before you go to the trouble of taking cuttings from it, because so many vines and trailing shrubs propagate easily from being layered. As you'll see from the directions below, it takes only a few minutes, a couple of tools, and some basic skills. But there is one thing that's indispensable – patience! You may have to wait a season or a year to see the results of your efforts.

Some plants root so easily from their branches that you need only bend them down and put them into contact with the soil. Others, however, are slightly more resistant. Assume that you'll have to go to a little trouble until you know the characteristics of the plant you're propagating.

Any stem you are layering into soil must stay in place while it's rooting, so as a first step, bend a few stems to the ground to see if they can flex without breaking. Once you've found the right stem (or stems), strip off the leaves on either side of the node you plan to bury, and wound the stem just below the node. If you place a matchstick, toothpick, or pebble in the wound, it will stay open and also form a callus that stimulates roots to form. For most plants, this is all that's required. If you don't have success, try dusting the wound with a little rooting hormone before you bury the stem.

Serpentine layering is similar, except you will bury two or more nodes of the same stem. Wound and bury alternate nodes, leaving every other node above the soil to continue photosynthesizing.

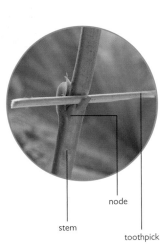

stem

node

toothpick

bent wire

leaves

Long-stemmed or trailing plants such as clematis can be serpentine layered by wounding and burying alternate nodes along the stem.

APPROPRIATE PLANTS

COMMON	BOTANICAL
Kiwifruits	*Actinidia* spp.
Serviceberries, Juneberries	*Amelanchier* spp.
Japanese Laurel	*Aucuba japonica*
Bougainvilleas	*Bougainvillea* spp.
Grape Ivies	*Cissus* spp.
Summersweets, Japanese Clethras	*Clethra* spp.
Winter Hazels	*Corylopsis* spp.
American Hazels	*Corylus* spp.
Smoke Trees	*Cotinus* spp.
Cotoneasters	*Cotoneaster* spp.
Daphnes	*Daphne* spp.
Heathers	*Erica* spp.
Euonymuses	*Euonymus* spp.
Forsythias	*Forsythia* spp.
Fothergillas	*Fothergilla* spp.
Ivies	*Hedera* spp.
Hibiscuses	*Hibiscus* spp.
Hoyas	*Hoya* spp.
Hops	*Humulus* spp.
Climbing Hydrangeas	*Hydrangea anomala* sbsp. *petiolaris*
Hollies	*Ilex* spp.
Mountain Laurels	*Kalmia* spp.
Privets	*Ligustrum* spp.
Sweetgums	*Liquidambar* spp.
Honeysuckles	*Lonicera* spp.
Magnolias	*Magnolia* spp.
Bayberries, Sweet Gales	*Myrica* spp.
Boston Ivies, Virginia Creepers	*Parthenocissus* spp.
Philodendrons	*Philodendron* spp., *Monstera* spp.
Pierises	*Pieris* spp.
Pittosporums	*Pittosporum* spp.
Rhododendrons	*Rhododendron* spp.
Currants, Gooseberries	*Ribes* spp.
Blackberries, Raspberries	*Rubus* spp.
Scheffleras	*Schefflera* spp.
False Spireas	*Sorbaria* spp.
Spiketail	*Stachyurus praecox*
Stephanandras	*Stephanandra* spp.
Stephanotis, Bridal Wreath	*Stephanotis floribunda*
Snowbells, Fragrant Styraxes	*Styrax* spp.
Lilacs	*Syringa* spp.
Glory Flowers	*Tibouchina* spp.
Star Jasmine	*Trachelospermum jasminoides*
Blueberries	*Vaccinium* spp.
Viburnums	*Viburnum* spp.
Periwinkles	*Vinca* spp.
Grapes	*Vitis* spp.

Any plant with long-enough stems can be serpentine layered, but this technique is particularly appropriate for the following plants.

Clematises	*Clematis* spp.
Pothos	*Epipremnum aureum*
Hops	*Humulus* spp.
Honeysuckles	*Lonicera* spp.
Philodendrons	*Philodendron* spp.
Blackberries, Raspberries	*Rubus* spp.
Grapes	*Vitis* spp.
Barren Strawberries	*Waldsteinia* spp.
Wisterias	*Wisteria* spp.

1 INVESTIGATE

Bend the stem down to the ground to see where it will make contact with the soil and whether you can do a serpentine layering with it.

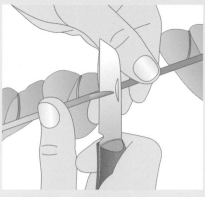

2 WOUND

Cut through the underside of the stem just below the nodes you are burying, about a third of the way through.

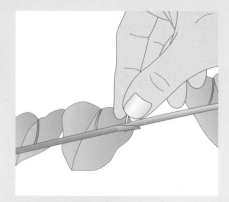

3 HOLD OPEN

Insert a wooden matchstick, toothpick, or small pebble to keep the wound open.

4 PIN DOWN

Use the ground staple or U-shaped wire to secure the stem. Push soil over it, and press down to exclude air pockets. Place a large rock over the buried portion to be doubly sure it stays in place. Keep the soil moist while the stem is rooting.

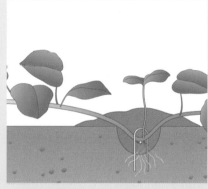

5 MONITOR

Eventually you'll see new growth on the stem, indicating that it has rooted. Wait until the new plant is growing vigorously before severing it from its parent.

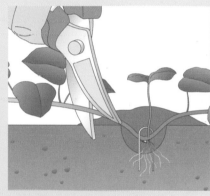

6 SEVER

Once the new plant is large enough, cut it from the parent and any other new plants. Leave it in place for a season, and then move it to its permanent position in the garden.

Checklist

Season: Spring

Tools: Trowel, spade, long-handled utility knife, bypass pruners

Equipment: Ground staples or bent 9-gauge wire, small wooden matchstick, toothpick, or pebble, small watercolour brush for rooting compound

Supplies: Rooting compound for the occasional plant

What can go wrong

Roots don't form: Be patient. Some plants take a year before forming roots, so give your plants time. The second cause of poor rooting could be lack of water or compacted soil. Make certain that the soil around the buried stem is friable and consistently moist.

Be careful to leave enough leaves above the soil so the plant can continue to photosynthesize.

Air layering

You may never need to consider air layering until a favourite houseplant threatens to touch the ceiling or drops so many bottom leaves that it has lost its good looks. Even then, you may be reluctant to carry out this procedure for fear of losing the plant. Fortunately, air layering is much easier than it looks and, when environmental conditions are correct, almost always foolproof.

The first essential to successfully air layering houseplants is cleanliness. If the medium you're rooting into is initially sterile, the plastic you use to cover the medium and stem is clean and fresh, and any other plants in the area disease free, your chances are good. Temperature and humidity are the remaining factors to consider; if the air is so cool or dry that the plant has a hard time maintaining its health, it won't be able to spare the energy to develop roots. In this case, wait until spring or summer for the temperatures to rise, and if the air is still dry, use a humidifier in the room while the plant is rooting.

Woody outdoor plants can also be propagated by air layering and require the same sort of consideration that you give to houseplants: good sanitation and an appropriate environment. But the technique is somewhat different – you remove a ring of bark around the stem, as shown on the facing page, before making the vertical cut into it.

slit —

toothpick

stem —

APPROPRIATE PLANTS

COMMON	BOTANICAL
Bougainvilleas	*Bougainvillea* spp.
Camellias	*Camellia* spp.
Citrus trees	*Citrus* spp.
Croton	*Codiaeum variegatum*
Ti Plant	*Cordyline fruticosa*
Dumb Canes	*Dieffenbachia* cvs.
Corn Plant	*Dracaena fragrans*
Rubber Trees, Figs	*Ficus* spp.
Gardenias	*Gardenia* spp.
Witch Hazels	*Hamamelis* spp.
Hibiscuses	*Hibiscus* spp.
Hollies	*Ilex* spp.
Magnolias	*Magnolia* spp.
Oleander	*Nerium oleander*
Azaleas	*Rhododendron* spp.
Scheffleras	*Schefflera* spp.

Checklist

Season: Spring for outdoor plants; spring or summer for houseplants

Tools: Sharp garden knife

Equipment: Small wooden matchstick or toothpick

Supplies: Unmilled peat moss or coir, construction-grade plastic sheeting, twist ties or garden twine, aluminium foil, rooting compound

What can go wrong

Roots don't form: Indoor plants should form roots in a couple of months, but woody outdoor species may take a year or more to root. Simply keep the peat moss moist and make certain it's not overheating, and your patience will likely be rewarded.

Completely enclose the peat moss in the plastic or it will dry before the plant has a chance to form roots.

Air layering indoor plants and monocots

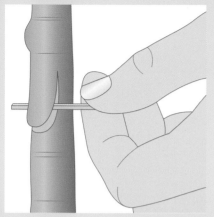

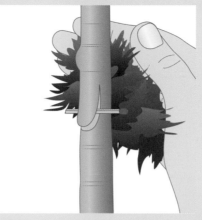

1 WOUND
After making a 2½- to 4cm slit about a third of the way through the stem in the current season's growth, use a wooden matchstick or toothpick to hold the cut open.

2 PACK
Soak the unmilled peat moss in warm water for an hour or more, until it is saturated. Squeeze it out and pack it around the cut. Ask a friend to wrap plastic sheeting around the stem while you hold the peat moss in place.

3 COVER
Use a twist tie or garden twine to secure the plastic sheeting to the stem at both top and bottom. If the plant is in a sunny location, cover the plastic with aluminium foil to prevent the stem from overheating under the plastic. Monitor to make certain that the peat moss stays moist. Roots should form in 6 to 8 weeks. Wait until they fill the enclosure before severing the newly rooted plant from the parent stem and transplanting it into a new pot.

Air layering a woody plant

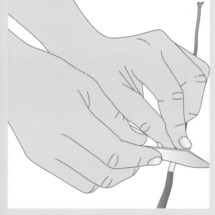

1 RING
Choose a stem about the width of a pencil from the previous season's growth. Just under a node, make two horizontal cuts about 2cm apart in the bark and cambium layer. Make one vertical cut, joining the two horizontal cuts, and pry off the bark and cambium layer from this area. Scrape off all of the cambium layer to expose the more solid-looking wood under it.

2 PACK
Dust the exposed wood with rooting compound. Saturate unmilled peat moss, squeeze it out, and pack it around the exposed wood. Ask a friend to wrap it in plastic sheeting while you hold the peat moss in place.

3 COVER
Secure the top and bottom of the plastic with a twist tie to hold the peat moss in place, then cover the plastic with aluminium foil to exclude light. Roots should form in about a month. Wait until they are 2½cm long before severing the newly rooted plant from the parent stem.

Stooling or mound layering

This unusual propagation method is most effective with shrubs that grow a great many stems in close proximity to each other. Because the crown of the plant is capable of producing so many stems, it can also produce many new plants at a time.

Commercial orchardists and nursery people very often stool plants to propagate particular rootstocks or nursery specimens because the method is fast, reliable and extremely productive. If you are starting an orchard and want to grow your trees on a particular rootstock, you might consider stooling your own. It will take a few years for the rootstocks to develop to grafting age, but you'll never doubt their identity or health.

This technique can also allow you to develop a striking hedge without spending a mint at the nursery. For example, if you wanted to put in a long hedge of mock orange bushes between your garden and the neighbour's property, the price for buying enough plants could be prohibitive, but it's not at all costly to buy only a few plants and stool them. It might take two more years until the hedge would be large enough to serve as a visual boundary, but it would certainly be less expensive and give you very uniform results.

APPROPRIATE PLANTS

COMMON	BOTANICAL
Aralias	*Aralia* spp.
Croton	*Codiaeum variegatum*
Hazels	*Corylus* spp.
Quince	*Cydonia oblonga*
Deutzias	*Deutzia* spp.
Lavenders	*Lavandula* spp.
Apple rootstocks	*Malus* spp.
Mock Oranges	*Philadelphus* spp.
Gooseberries, Blackcurrants	*Ribes* spp.
Blackberries, Raspberries	*Rubus* spp.
Spireas	*Spiraea* spp.

Checklist

Season: Mid spring

Tools: Spade, trowel, bypass pruners or garden knife

Supplies: Fully finished compost or nutrient-rich topsoil

What can go wrong

Parent plant dies: Remember to leave the top few centimetres of each stem unburied so that its leaves can continue to manufacture sugars to sustain growth.

leaves
stem tips
buds
stem

Buried too deeply, buds are covered and plants will die.

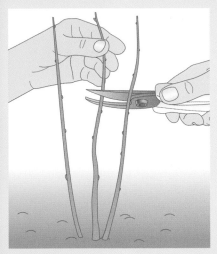

1 CUT

The first winter after planting a shrub you're going to stool, cut the central stems back to about 30 to 46 cm from the soil surface to stimulate more stems to form.

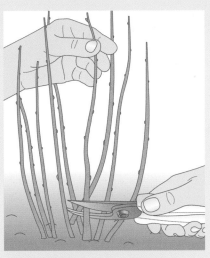

2 CUT AGAIN

The following year, in late winter or early spring, cut back all the stems on the plant to about 2cm above the soil surface.

3 MOUND

In spring, let the new shoots that form grow to about 15cm tall, and then begin mounding soil over them. Leave the top 5cm of the shoot exposed.

4 MOUND AGAIN

Add soil as the season progresses, until about half the stem is covered. Roots will form on the underground portion.

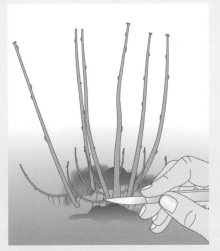

5 SEVER

The following spring, cut the rooted shoots from the parent plant and transplant them into a nursery bed where they can grow until they reach transplant size. Give them adequate room to develop strong root systems and straight trunks.

6 REPEAT

The parent plant will often produce a second set of shoots. Treat these as you did the first if you want more plants from the parent plant.

French layering

This technique is also known as trench layering, continuous layering, and etiolation layering. At first glance, it resembles serpentine layering (page 102). The difference is in the amount of the stem that is buried. In contrast to serpentine layering (in which you bury every other node along the branch), all nodes but the last two or three are buried. Consequently, it's used to propagate plants with stiff, woody stems rather than those that bend easily.

French layering is most frequently used by commercial growers to propagate rootstocks for various orchard crops. They begin by planting a row of the desired rootstock, leaving a metre or so between the plants in all directions. They then French layer the plants, ending up with a row of seedling-size rootstocks that they can separate and grow on until they graft onto them.

Few people with a back garden need the number of plants produced by French layering. However, if you are developing an orchard and want to use a particular rootstock or plant a uniform hedge of a plant that takes well to this technique, it's worth learning how to do it – it's an easy way to radically increase the number of plants. A well-cared-for mother bed can last for years, yielding successive crops of new rootstocks.

wire
stem

nodes

Checklist

Season: Early autumn and then early spring for 2 successive years

Tools: Spade, garden knife, bypass pruners

Equipment: Ground staples or U-shaped pieces of 9-gauge wire, wooden matchsticks or pebbles

Supplies: Fully finished compost or nutrient-rich topsoil, stones

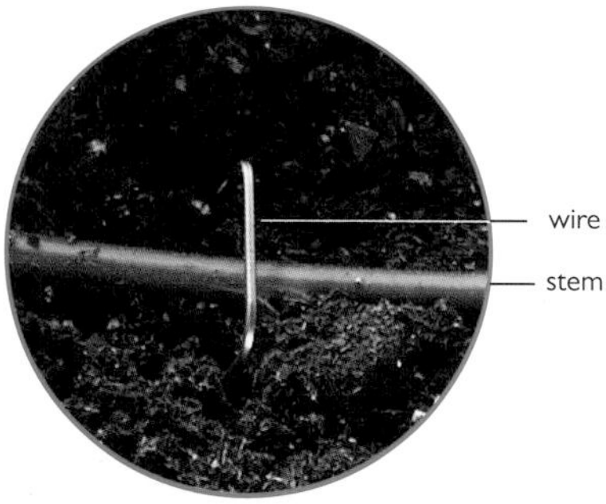

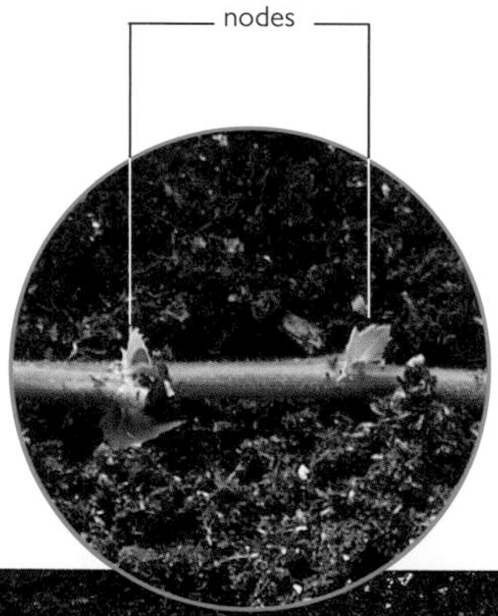

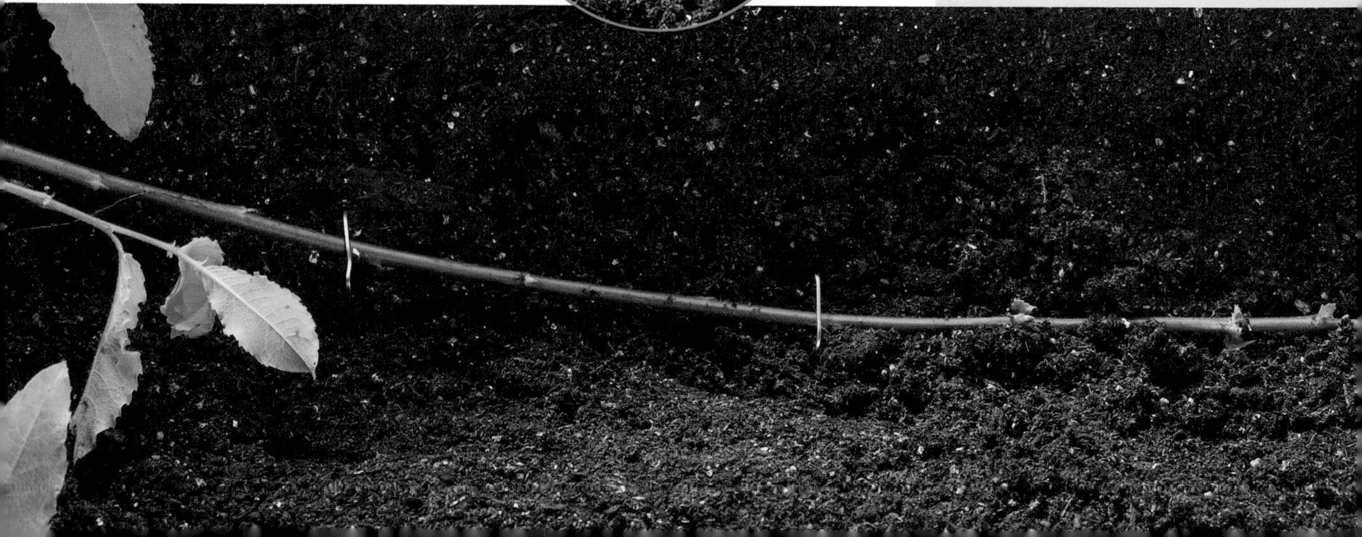

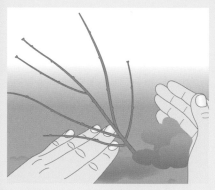

1 PLANT
Prepare for French layering by buying the plant you wish to propagate a season in advance. In the autumn, plant it at a 45-degree angle to the soil surface to increase growth to one side.

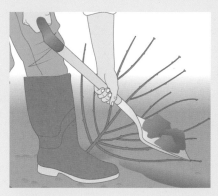

2 DIG
Early next spring, dig a trench into the soil where you'll bury the stems of the plant. If the trench soil does not drain freely, add several centimetres of sand and compost.

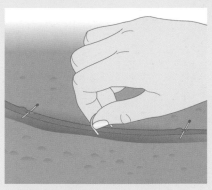

3 WOUND/SECURE
Strip leaves from the stems and make cuts in them, just below the nodes. Position a matchstick or pebble in each cut to hold it open. Then lay the stems in the trench, and secure them with ground staples or pieces of wire.

4 FILL/BURY
Roots and shoots will form over the season. Add soil over the shoots as they grow to encourage them to form more roots, and keep the soil consistently moist but not soggy.

5 DIG
The next spring, dig the stems carefully so you don't injure the new roots. Once the stems are exposed, cut them into sections, each with a single plant. Discard the stem sections between the new plants.

6 TRANSPLANT
Plant the new plants deeply in the soil of a nursery bed, leaving ample room around them so they develop strong root systems.

What can go wrong

Nodes don't root: Depending on the plant, rooting can take anywhere from a few weeks to a full season. If you have done everything correctly, be patient – your stem will probably root.

Stem rots: Plant in a friable, well-drained soil to protect the plant against fungal infections. If you have any doubt about the drainage capacities of the soil, dig the trench a few centimetres deeper than necessary, and fill it with a mixture of sand and fully finished compost.

Ensure the nodes of the rootstock stem are adequately pinned and buried in the prepared soil bed.

Dropping

Dropping is another propagation technique that home gardeners seldom undertake. Depending on the method you choose – there are two – it can involve a fair bit of heavy work, so it doesn't appeal to everyone. Beyond that, it can seem alarming – one of the two techniques depends on "dropping" the majority of your plant below the soil surface and burying it.

If you look at the small list of species suitable for dropping (above right), you'll notice that they are all extremely tough. Their native habitats are uniformly harsh – places where cold winds consistently blow, topsoils are thin, or flooding can put crowns underwater for protracted periods of time. This toughness is the quality that enables them to not only survive when they are dropped but also to respond by producing an enormous number of small plants.

As a home gardener, you can save a lot of money by learning this technique. If, for example, you want to grow a number of blueberry plants, you can

buy one or two of each of the cultivars you want and propagate them by dropping. Each plant can give you at least six to ten new ones – all at the cost of the original. So don't dismiss dropping as impractical until you've tried it; you may decide it's the most practical propagation technique you know.

APPROPRIATE PLANTS

COMMON	BOTANICAL
Scotch Heather	*Calluna vulgaris*
Irish Heath,	
St Daboec's Heath	*Daboecia cantabrica*
Heathers, Heaths	*Erica* spp.
Cranberry	*Vaccinium macrocarpon*
Blueberries	*Vaccinium* spp.

Checklist

Season: Mid spring or mid autumn

Tools: Spade, bypass pruners or sharp garden knife

Equipment: Ground staples or U-shaped pieces of 9-gauge wire

Supplies: Fully finished compost or nutrient-rich, friable topsoil

parent plant

stem tips

What can go wrong
Stems rot: Drop plants only in well-drained, nutrient-rich soils. Adding fully finished compost to the soil around the stems can help protect them from diseases because the compost contains beneficial organisms that prey on those that cause problems.

High labour, high return

1 PLANT
Buy and plant your stock plant a year before you plan to drop it. For best results, plant it in a loose, nutrient rich, friable soil.

2 DIG
In the second spring or autumn, dig the plant, and replant it in a hole that is deep enough so that all but a few centimetres of the stem tips will be buried when the soil is filled in around it.

3 MONITOR
As the stems grow, roots will be developing on their buried nodes. If you dropped the plant in spring, you can dig and transplant it as early as autumn of the same season; if you dropped it in autumn, wait until the following year. Check to see if the plants are ready in spring; if they're not, wait until autumn.

Less labour, lower return

1 PLANT
Buy a plant especially for propagating, or use one that is already growing. In either case, you'll need low-growing side branches to carry out this technique.

2 PEG
Peg low-growing branches on the perimeter of the plant to the soil surface, and then cover all but the tips with ½cm of soil. Roots will form within a month to 6 weeks.

3 SEVER
A year later, the roots should be sufficiently developed so the new plants can survive on their own. Sever them from the parent plant, and grow on in a nursery bed for a year or two before putting in permanent positions.

Beware mounding soil too much; it might be washed away and damage new plants.

GRAFTING

Grafting is one of the most exciting things you can do in the garden. Once you master it, you can create such things as an apple tree that contains two cultivars that pollinate each other, a mixed-cultivar tree that produces a variety of fruits in a small space, and specialty roses that grow on a trustworthy rootstock.

You can also do very practical things, such as develop an orchard at very little cost or order scion wood for a cultivar that is not locally available. Many heirloom fruits, for example, have great flavour but are no longer widely cultivated. For someone who knows how to graft, this is no problem; other growers can send you scion wood – branches or buds – of these old-fashioned fruits, and you can graft them onto appropriate rootstocks.

Rootstocks determine many of a plant's characteristics, including size, productivity, pest and disease resistance, fruit quality and stress tolerance. Consequently, there is no faster or more certain way to change or modify some basic qualities of the original plant. Perhaps you want to grow an apple that is notoriously susceptible to fire blight. By grafting a scion of the cultivar to a resistant rootstock, you'll have a much greater chance of being able to grow a healthy tree that never attracts this disease.

Similarly, if you want dwarf or semidwarf trees, you can grow the appropriate dwarfing rootstock and graft your chosen cultivars onto it. This allows you a great deal of flexibility – you might be able to grow three dwarf trees, each with a different fruit, in the space a standard would normally require.

Being able to graft can significantly shorten your wait for a particular flower or fruit, too. Some rootstocks speed up production of flowers or fruit, but beyond that, grafted plants automatically have a head start. As soon as a graft takes, the scion is growing on a plant with well-established roots. This means that the roots are capable of supplying the proper amounts of nutrients and moisture immediately, so you gain at least a year.

Thinking about all these advantages probably makes you want to start grafting trees and bushes.

Top left: Dicotyledonous cacti can be grafted.
Above: Graft citrus plants using a side-wedge or side-veneer graft.

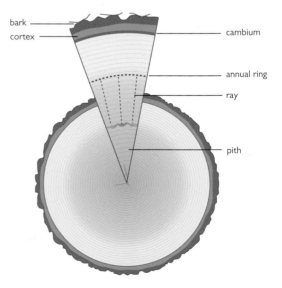

This enlarged cross section shows the interior of a stem.

Guidelines for success

• Choose compatible rootstocks and scions. Check with nurseries or an arboretum whenever you are in doubt about a possible combination.

• Raise your own rootstocks whenever possible. Use the stooling or French layering methods described on pages 106 to 109 to create enough rootstocks so that you can afford a failure or two.

• Grow your rootstock in nutrient-rich, deep, well-drained soil in a position with the correct environment – full sun, adequate moisture and possibly wind protection.

• Let your rootstock develop until it is about 46cm tall before grafting onto it, and remember to prune off all branches below 30cm to be flush with the stem in early spring. During the season, rub off any buds that develop on the stem.

• Choose your grafting technique depending on the relative size of the rootstock versus the scion, as described in the following pages.

• Use clean, sharp tools to make your cuts – sterilize your knife between cuts with a 10 per cent laundry bleach solution or another strong disinfectant.

• Make certain that the cambium layers of the rootstock and the scion touch each other before you tape or seal the graft with wax.

• Once you've made your cuts, work quickly – you don't want the wood to dry out.

• Make only one graft at a time; finish it before beginning on another.

• Don't touch the exposed wood – you could unknowingly transmit a disease.

• Use 15- to 30cm-long scions in warm, humid climates, but 15- to 25cm-long scions in cooler areas.

Successful grafting onto an established rootstock can result in a two-variety grapevine.

Propagate roses by bud grafting.

However, you may have heard that it is difficult and be reluctant to try for fear of failing. Give it a go anyway. In the first place, it's easier than it looks; in the second, nothing succeeds like failure. If you don't succeed with every graft at first, don't get discouraged. No one has a success rate of 100 per cent, and your failures can serve as teaching aids. The more you practise, the better you'll get. So take your grafting knife in hand, follow the guidelines above, and get going. You may surprise yourself with immediate successes – or simply learn the hard way how to achieve what you want.

Grafting with rootstocks

Grafting has been practised for a very long time. One of the first widely known written references was in 323 BC, which revealed grafting as an established practice in Greece. Chinese documents from a similar period also make it clear that it was common in the East.

Nature was probably the first teacher of grafting. Limbs growing in close proximity to each other may rub together in such a way that their cambium layers come into contact and form a natural graft. An observant fruit lover probably took over from here – experimenting and eventually producing better fruit by grafting a branch from a treasured tree onto the rootstock of a reliably vigorous variety.

Roses, apples, pears, peaches, plums, grapes, citrus, avocados and many woody ornamentals are all routinely grafted onto particular rootstocks, each of which confers certain characteristics to the resulting plant. Particular rootstocks are widely used in particular areas until they are supplanted by another that gives superior results. If you are considering growing your own rootstocks, refer to the charts at right and check with your local nursery to learn about the rootstocks considered the most reliable in your area. Find out if the rootstock you are considering is compatible with the scion(s) you plan to graft onto it.

You may be advised to graft an interstem between a standard rootstock and a scion to make up for incompatibility. Dwarfing rootstocks are either too shallow or too brittle to withstand extreme cold or freeze-thaw actions of soil. However, you can get around this problem by grafting a dwarfing interstem onto a standard rootstock. The standard rootstock survives the winter without a problem, and the interstem slightly dwarfs any scion that's grafted onto it.

Apple rootstocks

Apples are rarely grown on their own roots. The rootstocks listed below are among the most popular because they are dwarfing, offer resistance to a pest or disease problem, or stimulate earlier or heavier fruiting. Check with your supplier or a local fruit specialist to learn about their compatibilities.

ROOTSTOCK	CHARACTERISTICS
Malling 7 EMLA (M.7 EMLA)	Requires loamy soils to anchor well. Produces sturdy trees about 55 per cent the height of standards. Moderately resistant to crown rot and fire blight. Virus-indexed.
Malling-Merton 106 EMLA (MM.106 EMLA)	Semidwarfing. Confers early and heavy fruiting, moderate vigour, and resistance to woolly apple aphid. Cannot tolerate poorly drained soils because of susceptibility to crown rot. Virus-indexed.
Malling-Merton 111 EMLA (MM.111 EMLA)	Good anchorage; grows well in heavy and poorly drained soils but also has good drought resistance, so can be used in dry, sandy soils. Good choice for spur types. Tree height about 75 per cent of standard and averages 5½ to 6m. Resistant to woolly apple aphids; susceptible to mildew. Virus-indexed.
Budogovsky 9 (Bud. 9)	Trees grown on this rootstock average 25 to 35 per cent as tall as standards. Winter-hardy; resistant to crown rot; somewhat resistant to fire blight but susceptible to woolly aphids. Trees must be staked.
Malling 26 EMLA (M.26 EMLA)	Virus-indexed cross between M9 and M16 that produces trees about 45 per cent as tall as standards. Produces early fruit of high quality but susceptible to both crown rot and fire blight. Use it on well-drained soils and stake.
Malling 9 EMLA (M.9 EMLA)	Popular virus-indexed rootstock that produces trees only 35 per cent as tall as standards. Rootstock tolerates a variety of soil conditions, but roots can be brittle, so stake trees well, especially in areas with fluctuating winter temperatures.
Geneva 16	Developed by Cornell University in New York State; now recommended as a replacement for M9 because the roots are less brittle and create a better anchor. Resistant to fire blight and scab, but susceptible to woolly apple aphid and powdery mildew.
Geneva 30	Also from Cornell; replacement for M7 because the roots are better anchored and trees more productive. Resistant to fire blight and woolly apple aphid; somewhat tolerant to crown rot; susceptible to mildew.

EMLA indicates that the rootstock is resistant to several common virus diseases.

Pear rootstocks

Pears are often grafted to quince rootstocks, although Bartletts are sometimes used in commercial orchards in the northwestern United States. The OHxF series of rootstocks is a cross between Old Home and Farmingdale and is generally propagated by hardwood cuttings rather than stooling or French layering.

ROOTSTOCK	CHARACTERISTICS
Prv Quince BA29C	Dwarfing rootstock that produces trees about 60 per cent standard size, more drought tolerant, and higher yielding. Resistant to pear decline, crown gall and nematodes; susceptible to fire blight.
Quince A	Also dwarfing; offers resistance to pear decline, crown gall and mildew.
Bartlett Seedling	Has no effect on size but is hardy and vigorous. Used most frequently in the Northwest USA.
OHxF 513	Dwarfing rootstock that produces trees 3½ to 4½m in height, shortens time to fruiting, and grows well in most soils.
OHxF 87 (USPP#6362)	Patented rootstock; you cannot reproduce it, but its outstanding qualities make it worth mentioning. Like other OHxFs, it is resistant to fire blight, winter-hardy, well anchored, precocious and productive. More dwarfing than some; compatible with most varieties.

Rose rootstocks

Rose rootstocks are particularly sensitive about environmental conditions, so it's imperative to check with a local rose society before choosing one. The following rootstocks are the most commonly grown in the United States, Canada, Continental Europe, and the United Kingdom, as well as Australia.

ROOTSTOCK	CHARACTERISTICS
R. multiflora	Commonly grown in Canada and the northeastern, mid-Atlantic, and central United States; vigorous rootstock suitable for most types of roses. Poor choice for warm-winter areas because it requires winter chilling and is susceptible to root knot nematodes.
Dr Huey	Possibly the most popular rootstock in temperate areas. Used extensively because of its adaptability to various soil types and general vigour.
Manetti	Strong, tough rootstock resistant to nematode attack. Thorny, so can be difficult to handle when grafting onto it.
Fortuniana	Excellent rootstock for warm-winter areas. Has a deep root system that does well in dry, sandy soils. Strongly resistant to nematodes. Forms many suckers that must be removed if the scion is to prosper, but distinctive leaves make suckers easy to spot.

Peach, plum, apricot, almond rootstocks

All these fruits are in the *Prunus* genus, making them suitable for grafting onto many of the same rootstocks. But check with your supplier just to be certain—some combinations work better than others.

ROOTSTOCK	CHARACTERISTICS
St Julien	A semidwarfing rootstock frequently used for apricots, nectarines and peaches that will be trained to a fan-shaped trellis. Creates a bushy plum tree.
Brompton	Strong, vigorous rootstock compatible with all plum cultivars. Also used for fan-trellised nectarines and peaches but is not dwarfing.
Pixy	Semidwarfing rootstock used for plums that will stand on their own. Requires high soil fertility.

Cherry rootstocks

Cherries have traditionally been grown on Mazzard and Malhaleb rootstocks, but these are being supplanted by patented rootstocks such as Colt, Stella and the Gisela series.

ROOTSTOCK	CHARACTERISTICS
Mazzard F 12-1	Vigorous rootstock compatible with most sweet cherry varieties; gives good anchorage as well as some resistance to nematodes, bacterial gumosis and oak root fungus. Somewhat tolerant of the prune dwarf virus (PDV) and the *Prunus* necrotic ringspot virus (PNRSV), but susceptible to crown gall and bacterial canker.
Mahaleb	Like Mazzard, gives tolerance to PDV and PNRSV. Can produce a bushy habit but does not dwarf trees. Cold-hardy, drought-tolerant and productive. Resistant to bacterial canker and root lesion, but susceptible to oak root fungus, root knot nematodes and *Phytophthora*. Does not tolerate heavy soils.
Gisela® 12 (Gi 195-2) USPP #963	Semidwarfing and tolerant of a wide range of soil types, including heavy ones. Produces a precocious and productive tree that is resistant to a number of viruses. Must be staked.
Colt	Vigorous rootstock often used on cherries that will be fan-trained. Not very dwarfing; can be so vigorous that it inhibits fruiting unless the roots are pruned every few years.

Grape rootstocks

Wine grapes (*Vitis vinifera*) were grown on their own rootstocks until the grape phylloxera, from the eastern United States, was introduced to Europe in the 1860s. In 1868, this pest destroyed more than a third of the grapes in France and many vineyards in the rest of Europe, as well as in California. Growers experimented with using American grapes that resist these aphidlike insects as rootstocks, but many of them performed poorly in the high pH soils of France. Eventually, clonal selections of hybrids of *V. berlandieri* and *V. rupestris* came to be widely used, and today at least 30 are common in vineyards in the United States and Europe. If you are interested in growing your own grape rootstock, contact a local nursery to locate one appropriate for your area and the scions that you want to grow.

Whip-and-tongue grafting

If you are just learning to graft, this technique is a good starting point. It is fairly easy to execute, and the "tongues" you cut into the rootstock and scion wood make it easy to position the two pieces so the cambium layers are in contact with each other.

Whip-and-tongue grafting is a practical method for grafting fruit trees, particularly apples and pears, in most parts of the country. Not only is it a relatively simple graft to make, it also produces very straight trees. The only caution against this method is climatic: it works well in high-humidity areas, and it's not generally recommended for arid regions.

If you are planning to do whip-and-tongue grafting, you'll need to prepare a year in advance. In early spring, plant a 1-year-old rootstock where you want the tree to grow. Care for it throughout the year, rubbing off lateral buds as they form and keeping moisture and fertility levels adequate but not excessive.

If you plan to buy scion wood, arrange this during early to midsummer so that your supplier will be able to plan ahead. The supplier will send you the scion wood in early spring of the following year.

If you plan to cut your own scions, you'll need to do this while the branches are dormant. In very cold climates, late autumn is the best collecting time; midwinter is better in

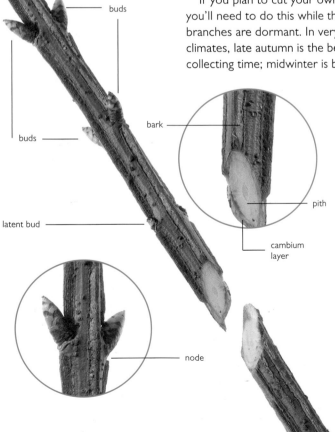

Apple and pear varieties are often propagated with whip-and-tongue grafting.

more moderate climates. Choose strong, straight branches — remember that they will eventually form the trunks of your trees. Cut 23- to 25cm pieces about 4cm below a bud, making certain that the width at the bottom of the piece is the same as that of the rootstock (ideally, about 1cm thick). Bundle the scions together with a rubber band or a twist tie before placing in moist peat moss or wrapping in a dry plastic bag. Label and keep them in a cool, dark area where they won't freeze. You can put them in a refrigerator, but don't store them with fruits or vegetables, which release ethylene that can prevent the buds on the scion wood from developing.

APPROPRIATE PLANTS
You can try this graft on almost any woody plant, but it is commonly used on fruit trees.

COMMON	BOTANICAL
Apple	*Malus* spp.
Apricot, Cherry, Nectarine, Peach, Plum	*Prunus* spp.
Pear	*Pyrus* spp.

buds

buds

bark

pith

latent bud

cambium layer

node

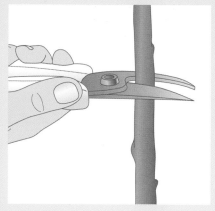

1 CUT STRAIGHT
Just before the buds break in early spring, cut off the top of the rootstock so it stands 15 to 30cm inches above the soil line.

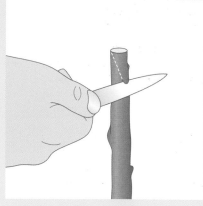

2 CUT OBLIQUELY
First cut the rootstock at an angle of 45 degrees to expose the maximum amount of cambium, then make a vertical cut through that area to make the "tongue".

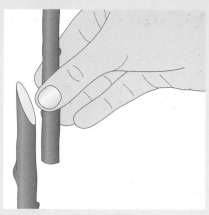

3 PLAN
Hold the scion up to the rootstock to plan where to make the cut. Make sure there's a bud on the scion, just above the cut.

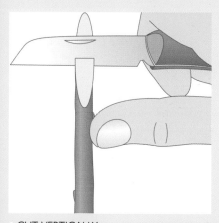

4 CUT VERTICALLY
After cutting the scion wood at a 45 degree angle to match that on the rootstock, make your vertical tongue cut. You only need three viable buds on the scion wood, so if it is too long, trim the top.

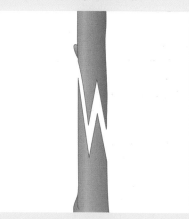

5 FIT TOGETHER
Fit the pieces together so that the two tongues form a lock. Wrap the union with electrical or grafting tape at least 2½cm above and below the area where the two pieces are joined together. In all but reliably humid areas, cover the tape with wax, and wax the top of the scion if you trimmed it. Label the plant, and note the work in your garden notebook.

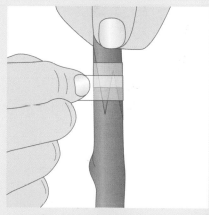

6 WAIT
Wait at least 2 months before removing the tape after a strong callus has formed and buds on the scion are developing. Don't let any buds on the rootstock grow; monitor carefully and rub off any that form, as well as suckers that arise from the soil.

Checklist

Season: Late winter to early spring – before flower buds swell

Tools: Pruning shears, grafting knife

Supplies: Alcohol for sterilizing the knife, grafting tape or electrical tape, grafting wax, labels, notebook and pen

Light: Cloudy

What can go wrong

Graft doesn't take: If either the rootstock or the scion wood dries out, the graft won't take. The union may not have been wrapped securely enough with electrical or grafting tape, or the two pieces of wood may have been dissimilar in size so that air penetrated some exposed wood. If your pieces are differently sized, cover the exposed wood with grafting wax after wrapping tape around the union.

Apical-wedge grafting and saddle grafting

Both of these techniques are relatively simple and a good choice if you are just learning how to graft and want to build your confidence. Commercial growers tend to use either the apical-wedge or saddle technique for woody ornamentals such as rhododendrons and witch hazel. You can do this at home, too, but you may have to search for a nursery that is willing to send you a rootstock. You can usually find good scion wood in the gardens of friends and neighbours.

Similarity of size is the most important factor when grafting with either of these techniques, which simply can't work unless the rootstock and the scion match almost exactly. Consequently, it's important to time this operation well or have a number of choices for your scion wood. Choose another grafting method if the cambium layers of your pieces won't knit together all along the cut.

Cleanliness is always important when you are grafting. Because gardeners work in the soil, they are sometimes lax about this aspect; after all, if rain doesn't splash soil onto the stems and leaves of plants, wind blows dust onto them. However, when working with newly exposed tissue, the picture changes. Regard this wood as you would an open, gaping wound on an animal or person. Not only does it lack skin – or bark – to protect it against pathogens, the moist, nutrient-bathed tissue makes an ideal breeding ground for any pathogen that lands on it. If you keep this in mind, it will be easier to refrain from touching the exposed wood with anything but the sterilized blade of your knife.

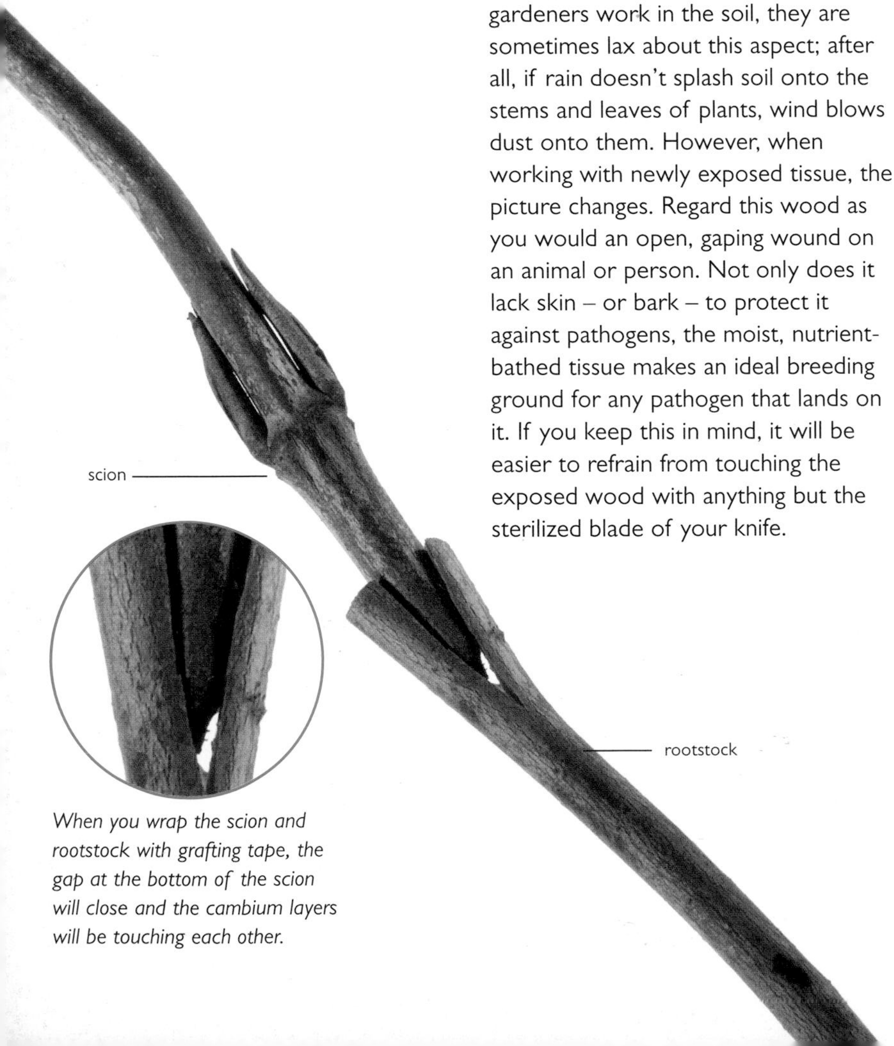

scion ———

rootstock ———

When you wrap the scion and rootstock with grafting tape, the gap at the bottom of the scion will close and the cambium layers will be touching each other.

APPROPRIATE PLANTS
This technique will work well with any woody plant that you want to grow on a rootstock other than its own. Commercially, it is frequently used for the following species.

COMMON	BOTANICAL
Oranges, Lemons, Grapefruits	*Citrus* spp.
Witch Hazels	*Hamamelis* spp.
Hibiscuses	*Hibiscus* spp.
Rhododendrons, Azaleas	*Rhododendron* spp.

Checklist
Season: Early spring, just as new growth is about to begin

Tools: Pruning shears, grafting knife or utility knife

Supplies: Alcohol for sterilizing blades, paper towels, grafting tape or electrical tape, grafting wax, labels, notebook and pen

Light: Cloudy

What can go wrong
Graft doesn't take: Not only must the rootstock and scion wood match in diameter, the angles at which they are cut must also match. Use straight (not curved) cuts, and press the two pieces together firmly but gently.

Hibiscus responds to apical-wedge grafting.

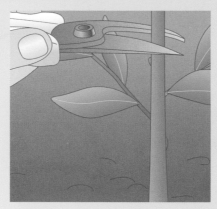

1 PLANT

Plant the rootstock a year before you want to make the graft. Place it in its permanent position or in a nursery bed with ideal growing conditions.

2 CUT

The following spring, cut off the top of the rootstock. Make the cut as straight as possible.

3 SLICE DOWN

Next make a vertical cut, about 4cm long, down through the stem of the rootstock.

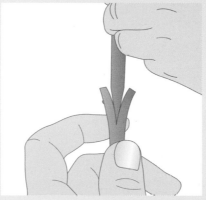

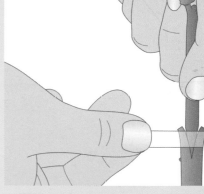

4 CUT SCION

Cut both sides of the bottom of the scion into a tapering wedge, or V shape, about 3cm long.

5 FIT

Insert the wedge of the scion wood into the slit you cut in the rootstock. Work carefully so you don't break off part of the rootstock.

6 WRAP

Once you are certain that the cambium layers of the two pieces are in contact with each other, wrap the union with grafting tape. Add a layer of wax if the union is not completely matching. Leave the tape in place until a callus has formed and the buds on the scion have started to develop.

SADDLE GRAFTING

Saddle grafting is the opposite of wedge grafting: An upward-pointing wedge is cut on the rootstock, and the scion is cut in an inverted V shape – the two pieces fit together as they do in wedge grafting.

Side-wedge grafting

Side-wedge grafting differs from apical-wedge grafting in two important ways: it is set into the side of the rootstock, not the top, and the top growth of the rootstock is allowed to continue growing while the graft becomes established.

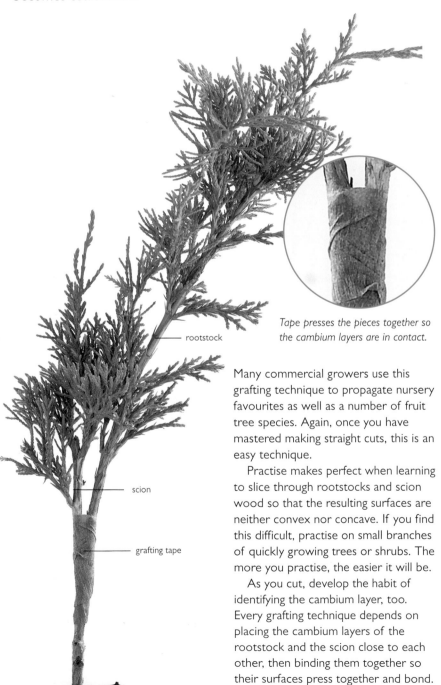

rootstock

scion

grafting tape

Tape presses the pieces together so the cambium layers are in contact.

Many commercial growers use this grafting technique to propagate nursery favourites as well as a number of fruit tree species. Again, once you have mastered making straight cuts, this is an easy technique.

Practise makes perfect when learning to slice through rootstocks and scion wood so that the resulting surfaces are neither convex nor concave. If you find this difficult, practise on small branches of quickly growing trees or shrubs. The more you practise, the easier it will be.

As you cut, develop the habit of identifying the cambium layer, too. Every grafting technique depends on placing the cambium layers of the rootstock and the scion close to each other, then binding them together so their surfaces press together and bond.

Rhododendron spp. may be grafted with a side-wedge graft.

1 PLANT

In early spring, plant a 1-year-old rootstock in an open nursery bed or in a pot that you can hold in a greenhouse or a cold frame over the first winter. Trim the bottom side shoots from the stem to prepare the area to accept a graft the following year.

2 CUT SCION

In later winter or early spring the following year, cut scions. They should be terminal shoots 10 to 15cm long and must have a terminal bud as well as a minimum of three lateral buds. If you cannot graft right away wrap the scions in plastic and keep them in a dark, cool place where they won't freeze.

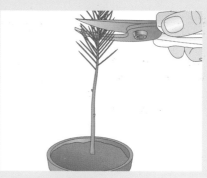

3 CUT ROOT STOCK

Make a straight cut across the top of the rootstock stem, about 20cm from the soil surface. Use a soft, damp cloth to wipe off the stem where you will make the graft.

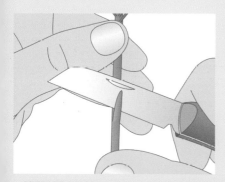

4 PREPARE SCION

In very early spring, remove the leaves or needles from the stem of the scion about 4cm up the stem, and then recut the base of the stem. Now make two 2½cm-long tapered cuts into the bottom of the stem—don't make the bottom a sharp point, but do make it narrow.

5 PREPARE ROOTSTOCK

Make a "flap" in the stem of the rootstock by making a 2½cm-long cut under the bark of the rootstock. Be certain to expose the cambium layer as part of the cut portion.

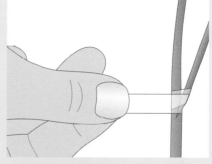

6 FIT TOGETHER

Fit the scion into the flap in the rootstock so that the exposed cambium layers in the scion are in touch with that in the rootstock. Tape the union. Unless the air is humid, cover the top of the plant, both scion and rootstock, with a plastic bag to maintain high humidity. Monitor to prevent excessive humidity or drying out, and remove the bag in about 6 weeks, after the graft has taken.

What can go wrong

Scion rots: Humidity levels must be high enough when you're grafting into the stem of a plant so that the tissues do not dry out. Tape and wax both retain moisture, but directions for some plants, such as the conifers shown above, ask that you also cover the graft with plastic film. Monitor carefully – it's as easy for a graft union to rot as to dry out.

Scion doesn't take: The rootstock and scion wood must be compatible. To locate an appropriate rootstock for the plant you wish to grow, check with a local expert or a knowledgeable nursery that grafts its own stock.

Checklist

Season: Late winter to early spring

Tools: Pruning shears, grafting knife or utility knife

Supplies: Alcohol for sterilizing blades, paper towels, grafting tape or electrical tape, plastic bag, grafting wax, labels, notebook and pen

Light: Cloudy

Approach grafting

Approach grafting is one of the most foolproof techniques because both the rootstock and scion grow on their own roots and have their own top growth while the graft forms. This minimizes the chances of rot or drying out, and both plants are supplied with adequate nutrition and water. This technique is the easiest for children to learn – they can readily see what's happening, plus success is more certain than failure.

APPROPRIATE PLANTS	
COMMON	BOTANICAL
Hollies	*Ilex* spp.
Apples	*Malus* spp.
Cherries	*Prunus* spp.
Peaches	*Prunus* spp.
Plums	*Prunus* spp.
Pears	*Pyrus* spp.
Tomato	*Solanum lycopersicum*
Blueberries	*Vaccinium* spp.
Grapes	*Vitis* spp.

Approach grafting is an excellent technique to use when you are trying to avoid soil-borne problems. For example, if you were growing a type of tomato that was susceptible to a soil-borne disease but knew of a tomato that resisted it, you could sidestep the problem by grafting your chosen tomato to the resistant rootstock. Similarly, the soil in some areas is so acid or alkaline that many cultivars have difficulty surviving even if the species itself tolerates these conditions. But if you approach graft the cultivar to the stronger species, the plant will probably thrive.

On plants such as hollies, grapes and blueberries, this method allows you to graft more than one gender or cultivar onto the same rootstock without running the danger of killing the rootstock. As long as the two stems you want to graft together are the same size and the current year's growth, the rootstock plant can be of any age. With blueberries and hollies, this means that you can make your own "self-pollinating" plants; with grapes, you could grow a mixed crop on every vine.

grafting tape

scion

root stock

Grafting tape holds the scion and rootstock in contact with each other.

1 PLANT
In very early spring, plant the "scion" that you want to graft to a rootstock in a small pot. The scion could be a cutting you've taken from another plant or a plant you've purchased.

2 CUT SCIONS
In later winter or early spring the following year, cut scions. They should be terminal shoots 10 to 15cm long and must have a terminal bud as well as a minimum of three lateral buds. If you cannot graft right away, wrap the scions in plastic and keep them in a dark, cool place where they won't freeze.

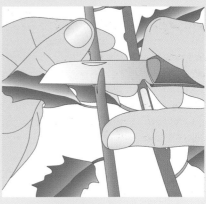

3 EXPOSE CAMBIUM
Check to see where the stems will touch and that they are the same diameter at that point, and remove a 3½- to 5cm piece of bark from both.

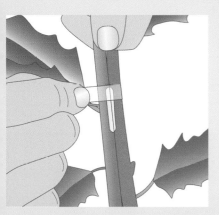

4 TAPE
Make certain that the cambium layers are in contact with each other, and then tape the union together.

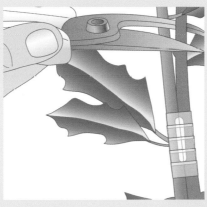

5 CUT BACK
Cut back the upper portion of the grafted stem on the rootstock plant to force it to send more nutrients to the area where the graft is healing.

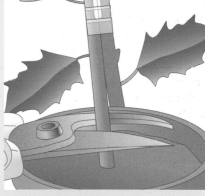

6 TRIM
After about a month to 6 weeks, the graft will have taken. On either side of the new graft union, cut off the roots from the scion plant and the top of the stem from the rootstock plant.

Checklist
Season: Spring to early summer

Tools: Grafting knife or utility knife

Supplies: Alcohol for sterilizing blades, grafting tape or electrical tape, labels, notebook and pen

Light: Cloudy

What can go wrong
Plants die: If you are grafting tender stems that lack any bark, be careful not to injure them by wrapping the grafting tape too tightly. It must be tight enough to hold the plants together, but not so tight that it crushes the vascular system of either stem.

Four-flap grafting

The four-flap graft looks alarming. But in this case, looks truly are deceiving. This graft is easy to execute and has an extremely high success rate. If you do it in the spring, you won't have to be a master cutter to strip away some bark from the rootstock, cut off the stripped portion of the rootstock, insert the scion inside the bark strips and tape it all together.

Timing is important. Collect 15 to 20cm of scion wood from the previous season's growth between February and March—while the tree is still dormant but not long before spring—and store it at a temperature of about 2°C until you're ready to make the graft. Wrap the scions in damp paper towels or peat moss so they don't dry out. When collecting, look for straight growth—the scion will be the trunk of your new tree—with at least three plump buds above the bottom 5cm or so. The scions should be the same size or a tiny bit larger than the rootstock—ideally, about 1 to 2½cm in diameter.

You'll need to make this graft after the tree has broken dormancy but before the buds have opened—the usual recommendation is when the buds are about 2½cm long. But remember that bud size can vary, depending on the type of plant. The important thing is to make it late enough so that the bark easily peels away from the inner wood—test this on branches of nearby trees of the same species to be certain of your timing.

The sap may have begun to run a little when you make the cut on the rootstock. Grafts don't take well when bathed in sap, which is why most grafts are done while the tree is dormant. To prevent problems, cut your rootstock in two stages. Take the first cut 2½cm above the point at which you'll want to make the graft. Wait a couple of days, and then make the second cut, positioned where you want to work. The sap should have stopped flowing much by this time.

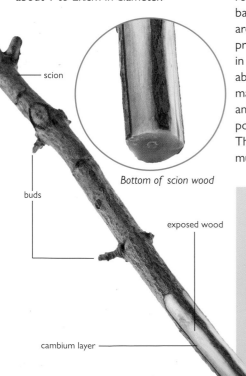

scion

buds

exposed wood

cambium layer

Bottom of scion wood

What can go wrong

Scion dies: If the scion dries out while the graft is forming, it will die. Wrap it as described in the illustrations to the right.

Both portions die: To prevent death from overheating, cover the graft with aluminium foil (as shown at right), and exclude air from the plastic bag you use to encase it.

Checklist

Season: Late winter for collecting scion wood; spring for grafting

Tools: Bypass pruning shears, grafting knife or sharp paring knife

Supplies: Alcohol for sterilizing blades, labels for scions, budding tape, aluminium foil, rubber band, grafting wax or white glue, freezer-weight ½-litre-size plastic bag, string

Environmental conditions: Cool, cloudy

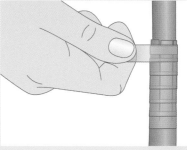

7 TAPE
Use budding tape to secure the bark strips in place; it is sufficiently elastic to allow the graft union to form under it. Make a complete seal with the tape to retain moisture, and then cover the tape with aluminium foil, shiny side out. This will help to keep the graft cool while it forms. Dab the top of the scion wood with white glue or grafting wax to retain moisture.

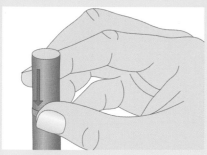

1 BAND ROOT STOCK

After making a straight cut across the rootstock, twist a rubber band onto it. Roll the rubber band about 5 to 7½cm from the cut surface.

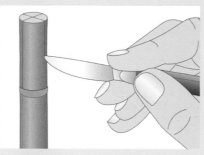

2 SLICE BARK

Mark an X on the top of the rootstock, and then make four cuts about 5 to 7½cm long just into the bark, using the arms of the X as a guide to make the cuts straight and equidistant from each other.

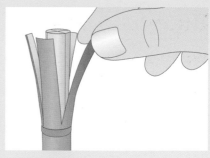

3 PEEL BARK

Peel back the bark, exposing the white wood underneath. After you've peeled each section, use the rubber band to hold the bark in place while you prepare the scion.

4 PREPARE SCION

Starting about 5cm from the bottom of the scion, cut away four pieces of bark, leaving thin strips of bark between your cuts. The resulting scion will be square rather than circular.

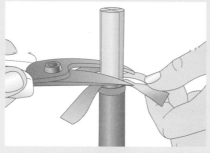

5 CUT ROOT STOCK

On the rootstock, cut off the exposed wood under the strips of bark. Make your cut as straight as possible.

6 FIT TOGETHER

Place the scion on top of the rootstock, positioning it so that the small areas of remaining bark are between the gaps formed when you draw up the bark from the rootstock. Use the rubber band to hold the bark strips in place while you work. Throughout, remember not to touch the exposed wood or inner surface of the bark strips with your fingers.

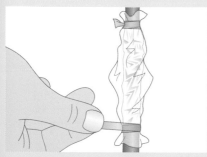

8 COVER

Cut the corner off a ½-litre-size, freezer-weight plastic bag and slip it over the graft. Secure it just under the aluminium foil, and push all the air out before securing it at the top of the graft. Don't let it extend beyond the foil or hold air – you don't want trapped air to heat and kill the graft.

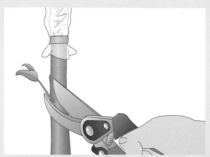

9 TRIM

Through the first season, buds on the rootstock will continue to grow. Prune off all but a few, and cut their tips back. If their growth becomes excessive, the growth of the developing branches on the scion will be inhibited. After a year or two, prune off all the branches on the rootstock and let the scion take over.

10 REMOVE COVER

After the scion is growing well, cut off the plastic bag and remove the aluminium foil. Cut through the tape to allow the graft to continue to expand. Rewrap the graft with the plastic bag, cover the bag with the aluminium foil, and tie the foil in place. Continue to check throughout the season to make sure the wood is not being strangled by the tape or ties. Remove all coverings the following spring.

Side-veneer grafting

Side-veneer grafting is a relatively easy technique. You do it on fairly small plants, so you can hold the scion up against the stem of the rootstock to determine where to make cuts that will match each other. Ideally, the scion and rootstock will be the same diameter, but this method also works if the scion is a bit smaller.

Because side-veneer grafts can easily be moved out of place, this technique is commonly used on potted plants in a greenhouse or other protected structure, so the graft won't be rubbed against by passing animals or subjected to strong winds. This graft was once used almost exclusively for camellias and rhododendrons that were resistant to grafting, but gardeners today use it for all types of young conifers, relatively small hardwood trees and shrubs, and broadleaf evergreens such as mangoes and avocados.

As with many other types of grafting techniques, you'll need to prepare your rootstock a year in advance. Plant it in a 4-litre nursery can, and care for it well during the first season. If it's a plant that requires a period of dormancy, leave it outside for at least 8 to 10 weeks during the winter months. If it is a tropical plant such as a mango, let the soil dry somewhat in autumn, and place it in a shady position for the winter. By late winter, it will be ready to use for grafting stock.

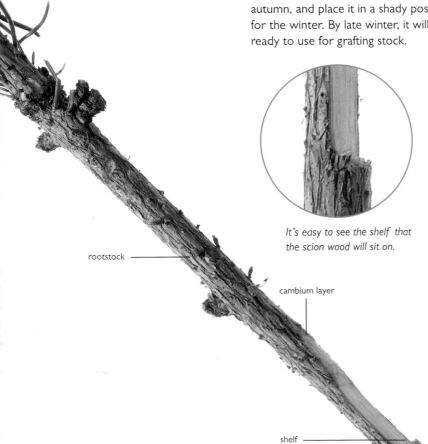

It's easy to see the shelf that the scion wood will sit on.

rootstock

cambium layer

shelf

APPROPRIATE PLANTS

COMMON	BOTANICAL
Firs	*Abies* spp.
Maples	*Acer* spp.
Birches	*Betula* spp.
Camellias	*Camellia* spp.
Junipers	*Juniperus* spp.
Mangoes	*Mangifera indica*
Avocados	*Persea americana*
Spruces	*Picea* spp.
Pines	*Pinus* spp.
Rhododendrons	*Rhododendron* spp.

Checklist

Season: Late winter

Tools: Grafting knife

Supplies: Alcohol for sterilizing blade, rubber bands, grafting tape, grafting wax

Environmental conditions: Cool, cloudy

What can go wrong

Scion dies: If the scion grows too much while the graft is forming, it will die. Keep the soil at about 18° to 24°C, but air temperatures at 7° to 10°C, for a month to 6 weeks after making the graft. Mist the plant frequently, but water it only when the soil is dry, and then only enough to keep it alive. In high light areas, shade the plant so that it does not lose a lot of water through transpiration.

Scion dies: The top growth on the rootstock helps to keep the scion alive during the 6 weeks or so that the graft is forming, so it's important to leave it in place until the scion is fully established. After the initial 6 weeks or so, the callus will have formed, and you can remove the grafting tape and begin to harden off the plant by exposing it to more light and higher temperatures. Wait an additional 6 weeks before cutting off the top growth of the rootstock.

1 TRIM ROOTSTOCK

Remove the bottom growth on the stem of the rootstock.

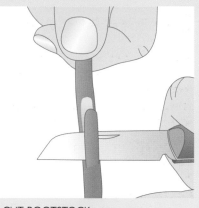

2 PLANT GRAFT

Hold the scion wood against the stem of the rootstock to determine the point at which the two stems are the same size. Mark the top and bottom of this area.

3 CUT ROOTSTOCK

Make your cut into the rootstock so that the scion wood will be at about a 30-degree angle when the two stems are bound together. Cut from the top down, but don't pull the cut portion off the stem at this point.

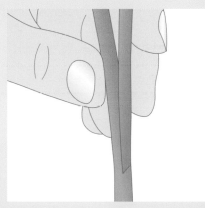

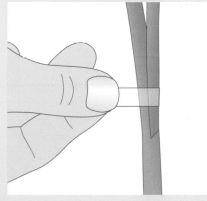

4 MAKE SHELF

Make a horizontal cut into the bottom of your previous cut. This will form a "shelf" for your scion to sit against when you bind the two.

5 FIT PIECES

After cutting the scion wood at the same angle that you cut the rootstock, place the two stems together.

6 WRAP

Wrap the graft carefully so that the cambium layers are pressed together. Cover the tape with grafting wax to keep the wood moist. Cut back the rootstock 2 inches above the graft. When the scion is actively growing, cut it back.

Graft Abies trees using side-veneer or side-wedge grafting in early to mid-spring. Ensure the graft stays in place until it has formed.

T-budding

T-budding – sometimes called shield budding – is a favourite technique of rose growers and both home and commercial fruit-tree growers. It is the most commonly used budding technique and offers some advantages for the beginner: you bud graft in early summer or autumn, when the buds on the scion wood, or bud stick, are fully developed, so you don't have to worry about storing scion wood. Second, if a bud doesn't take, your rootstock will barely notice the disturbance—you can try again with a new bud with no difficulty.

Budding is distinguished from other types of grafting because it is done with a single vegetative bud, not a length of stem. However, like other techniques, it lets you grow one cultivar of a particular plant on a rootstock that will give it characteristics such as dwarfing or disease resistance.

As with many other forms of grafting, you'll need to prepare a season in advance. In the spring or autumn of the year before you plan to make your graft – depending on your climate – plant the rootstock where you want your tree to grow for a few years after budding (in a nursery bed or its final position). It must be about as

thick as a pencil when you bud graft it.

Your bud stick should also be the thickness of a pencil. Take a vigorously growing, healthy shoot from the sunny side of the tree or bush. To keep it moist, place it in a bucket of water as soon as you cut it.

The directions at right are for T-budding shrubs and trees in almost every temperate climate. But if you are budding a citrus, or if you live in a tropical or subtropical area where the humidity is consistently high, you'll want to invert the T-cut so that the horizontal cut is at the bottom, not the top. This helps prevent the graft from being too moist.

APPROPRIATE PLANTS

COMMON	BOTANICAL
Apples	*Malus* spp.
Apricots, Cherries, Plums, Nectarines, Peaches	*Prunus* spp.
Pears	*Pyrus* spp.
Roses	*Rosa* spp.

Checklist

Season: Early summer in warm climates, after buds are fully developed and the bark easily lifts from the plant; fall in climates with dry summers. In all climates, water the rootstock well for 2 weeks before budding.

Tools: Budding knife

Equipment: Bucket of water

Supplies: Alcohol for sterilizing blade, grafting tape

Environmental conditions: Cool and cloudy; high humidity, if possible

What can go wrong

Bud doesn't take: It's important that your rootstock and scion bud are compatible. Check with a good nursery or plant association to determine appropriate rootstocks for your bud.

Bud dies: If you leave too much wood under the bud, it will interfere with the speed with which the bud takes. Be certain that you are taking only a sliver of wood under the bark and that the cambium shows. If you have too much wood, scrape it off before inserting it into the graft.

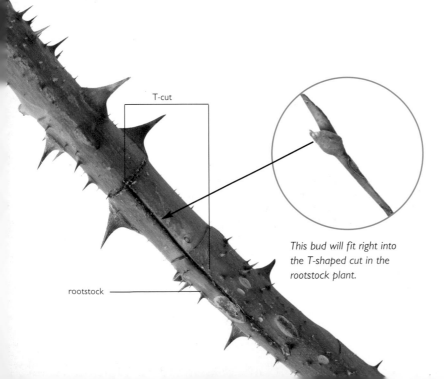

T-cut

rootstock

This bud will fit right into the T-shaped cut in the rootstock plant.

The top bud has too much wood under it to take well, while the bottom one has a good chance of taking.

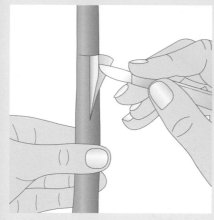

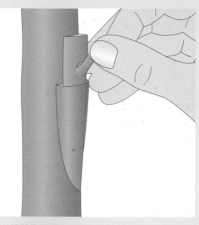

1 CUT "T"

Make your T-cut just into the bark of your rootstock in an area where there are no other buds. The cut should be 2½ to 7½cm above the soil surface. Make the vertical cut 2½ to 4cm long and the horizontal cut ½ to 1cm long. Use the end of the budding knife to pry open the two flaps of bark.

2 TRIM

After choosing a plump, healthy bud, cut off the leaf that is growing just under it, leaving a stub of the petiole that you'll be able to use as a handle. Make your first cut about 1cm below the bud, and slice into the woody tissue under the bud before drawing your knife upward. Angle your cut so that the knife emerges from the stem of the bud stick about 1cm above the bud. Hold the petiole to remove the bud. Turn it over to examine it, and scrape away woody tissue from under the bud, leaving the cambium and the bark.

3 INSERT BUD

Still holding the bud by the petiole, insert it into the T-cut, making certain that the lower end is positioned at the bottom. If the bark around the bud is larger than the T-cut itself, cut it off – it's important that the bud piece be the same size as the T-cut.

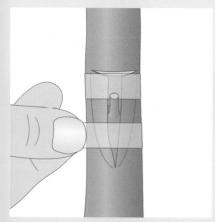

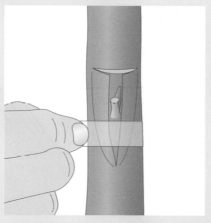

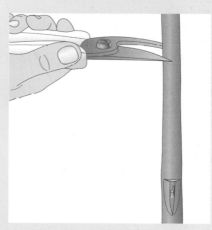

4 TAPE

Use grafting tape to bind the bud in place. Be careful to place the cambium layers together and not to cover the bud itself. Make the binding uniformly tight so that both the top and the bottom of the bud shield and the flaps of the rootstock are held in place.

5 LET GROW

Leave the grafting tape in place for about 2 weeks. After this time, check to see if the bud is taking. The petiole may have shrivelled, but the bud should be firm and plump. Carefully unwrap the grafting tape if the bud looks healthy and you can see that a callus has formed. Note: If you make the T-bud in autumn, wait until the following spring to unwrap the tape.

6 CUT

When you remove the grafting tape, you'll want to stimulate the growth of the bud. Make a cut halfway through the stem of the rootstock, about 10cm above the bud, and bend the top of the rootstock over. Once the shoot from the bud is about 8cm tall, cut off the top of the rootstock near the bud union.

Chip budding

Chip budding has traditionally been practised only by professionals – primarily nursery people propagating roses and fruit trees. However, this has changed in recent years. Home gardeners who tried chip budding discovered how much more forgiving it is than T-budding, and since then the word has spread. If you're interested in developing an orchard of heirlooms or specialty roses, try chip budding rather than T-budding.

Chip budding differs from T-budding in two important ways: first, you use a larger slice of wood; second, you can use this technique any time between midsummer and early autumn. You cut the bud differently, as well. Otherwise, the techniques are similar.

Again, take your bud stick from the sunny side of the tree, and choose only a vigorously growing, healthy shoot. When selecting a bud, make certain that it is a leaf bud, not a flower or fruit bud. You can tell the difference by looking – leaf buds are more pointed than flower buds and smaller and less rounded than fruit buds. To keep the bud stick moist, cut it right before you plan to make the graft, and immediately drop it into a bucket of water until you are ready to cut it.

Grafting tape holding the chip in place

other nodes on the rootstock

rootstock

APPROPRIATE PLANTS

COMMON	BOTANICAL
Hawthorns	*Crataegus* spp.
Magnolias	*Magnolia* spp.
Apples	*Malus* spp.
Apricots, Cherries, Plums	
Nectarines, Peaches	*Prunus* spp.
Pears	*Pyrus* spp.
Roses	*Rosa* spp.

bud

chip

Checklist

Season: Midsummer to early autumn

Tools: Budding knife

Equipment: Bucket of water, saucer of water

Supplies: Alcohol for sterilizing blade, wet paper towels, grafting tape

Environmental conditions: Cool and cloudy; high humidity, if possible

What can go wrong

Bud doesn't take: The cambium layers of the rootstock and bud wood must be in contact for the chip bud to take. If the diameter of the bud is just a little smaller than that of the rootstock, match the wood on one side so that the cambium layers here will be in contact with each other. If you place a smaller bud squarely in the middle of the rootstock, the graft will fail simply because the cambium layers don't touch on either side.

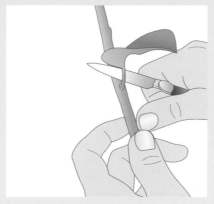

1 TRIM ROOTSTOCK

Choose a smooth area of wood on the shady side of the rootstock tree, several centimetres above the soil surface, where you will insert the chip bud. Remove branches and developing nodes from the stem that surrounds this area.

2 REMOVE LEAF

After you have chosen your bud, cut off the leaf just under it, but leave ½ to 1cm of the petiole to use as a handle.

3 CUT CHIP

Cut the chip from the bud stick in two steps. First cut about ½cm below the bud, making your cut at a downward angle of about 45 to 60 degrees and about ¼cm into the wood. Make your second cut about 1cm above the bud, about ¼cm into the wood and downward, too. It will meet the first cut under the bud. Holding the bud by the leaf petiole, drop it into the saucer of water to keep it moist while you prepare the rootstock.

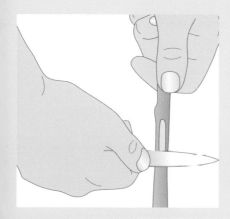

4 CUT ROOTSTOCK

Prepare the rootstock by cutting into the bark just above a node. Make this cut in two steps, first making an angled cut at the bottom as you did with the bud, then making the top cut. Make the cuts the same size, and leave a shelf for the bud to slip onto.

Bud dies: Because the sliver of wood is larger with chip budding than with T-budding, excessive drying of the bud is possible. Be certain to wrap the union completely to force the two pieces of wood tightly together and prevent air from getting in between and drying out the bud.

If the chip is this much larger than the rootstock, the cambium layers won't come into contact with each other.

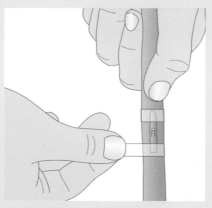

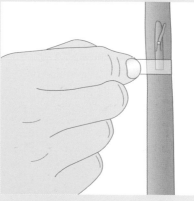

5 WRAP BUD

Wrap the bud tightly enough to hold it in place, but do not cover the bud itself. Remember to cover the petiole, too.

6 LET GROW

In about 2 months, the bud should have formed a callus and be growing as a part of the tree. Wait until the growth is strong before cutting off the top of the rootstock.

Grafting a cactus

You may never have considered grafting one cactus onto another. However, if you do begin to grow these plants, you may discover that you want to create a specialty collection filled with exotic species and vividly coloured blooms. You can buy these beauties, of course, but you can also save a lot of money by grafting your chosen plants onto appropriate rootstocks.

Choose a rootstock for your cactus according to the size of the scion you want to graft onto it as well as the climate in which you live. You may also consider whether or not the rootstock produces lots of offsets because you'll have to prune these off as they appear. If you ignore them, they can easily overwhelm the plant you've grafted onto the rootstock, defeating all your careful work.

Only dicotyledonous cacti can be grafted; monocotyledonous cacti such as aloes (*Asphodelaceae*) and agaves (*Agavaceae*) don't graft well because their vascular systems (where the cambium tissue is located) are randomly scattered throughout the stem. There are other considerations, too. For example, the more closely related the two plants, the better your chances of success. Success is also better assured if both plants are in an active growth phase and about the same size.

Check with local cactus societies to learn about rootstocks commonly used in your area. In general, *Echinocereus triglochidiatus, Echinopsis pachanoi,* and *Opuntia* spp. are commonly used for cold hardy cacti, while *Harrisia jusbertii* is used for slender cacti from warmer areas. *Nopalea* spp. are known for having fairly good tolerance to high humidity, warm temperatures and nematodes – a consideration if you are growing cacti in your garden soil. *Myrtillocactus geometrizans* is one of the best rootstocks for indoor plants and readily available from cactus suppliers.

Checklist

Season: Spring to early summer

Tools: Sharp garden knife, razor blade, or box cutter

Equipment: Clean 7½-to-10cm pot

Supplies: Alcohol for sterilizing blade, rubber bands, cactus soil mix, vermiculite, heavy gloves or tongs

Environmental conditions: Cool, cloudy; not too humid

What can go wrong

Scion takes and then dies: It's relatively easy to disrupt a delicate union between a rootstock and a scion when it is still young. Unless absolutely necessary, don't move the plant until several months after the graft has taken.

Scion rots: Rot is generally caused by excessive humidity. If ambient humidity is high, keep a fan blowing in the vicinity of the cactus to keep the air circulating.

scion

rootstock

A newly grafted cactus may fail if moved.

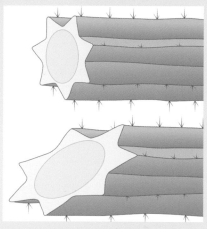

1 CUT

Use an extremely sharp knife, razor, or box cutter to slice off the top of the rootstock plant. Do not discard the top; instead, plant it in a pot filled with a cactus mix and topped off with a layer of vermiculite. It will root here, giving you another rootstock for future grafting efforts.

2 EXAMINE

Locate the vascular bundle in the center of the root. The cambium layer is the outermost ring that you see, and the area that must match when you make the graft.

3 TRIM

Cut off the edges of the rootstock at an angle, as shown above. Do not cut into the cambium layer, and do not touch the exposed tissue with anything but the sterilized knife blade.

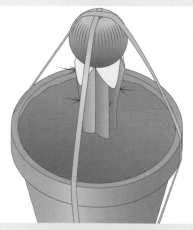

4 CUT SCION

Cut off a section of the plant that will be the scion, matching the diameters of the two pieces so the cambium layers will touch. Be careful to cut cleanly and without pressure – you don't want to injure the delicate cambium layer. Once again, don't touch the exposed interior with anything other than the knife blade.

5 PLACE

Place the scion on top of the rootstock, and gently rotate the scion back and forth a few times to exclude air bubbles.

6 SECURE

Use rubber bands to hold the two pieces together, and leave the pot in a well-lit area that's neither in bright, direct light nor too warm or humid. After 2 to 3 weeks, the union should have formed. If it has, cut the rubber bands so they fall away – do not pick up the plant to remove them. Leave the plant in place for another few weeks before moving it.

PLANT DIRECTORY

The following pages present a selection of plants to propagate, organized alphabetically by genus and supplying recommended propagation methods and potential problems. Tips help you get the most out of each technique. Choose the easiest method or others, depending on your interests and the circumstances of your garden. Check the handy icon strips to see at a glance which techniques you can choose from and which plants are appropriate for your hardiness zone (see page 183 for a zone map).

KEY TO SYMBOLS

seed   root cutting   leaf cutting   grafting

 division   stem cutting   layering

Abies spp.

Firs
PINACEAE

Zones: 4–8

The nearly 50 species of this genus come from North America, Europe, Asia and North Africa. They are characterized by needles that grow in a whorled pattern and are generally flattened. Needles range in colour from mid- to dark green, although species such as White Fir (*A. concolor*) have white or blue-green needles and others, such as 'Golden Spreader' (*A. nordmanniana*), have golden needles. Many species, including *A. veitchii* and *A. vejarii*, have needles with silver or gray markings on the undersides. Female cones are purplish blue to brownish and stand erect in the upper branches, while the purplish or brown male cones in the tree's crown hang down.

PROPAGATION METHODS

Easiest: *Seed.* Seeds ripen and are released from the female cones in autumn, and it's best to plant them immediately. Soak them in warm water for about 24 hours, and then plant them in deep trays or pots and place them in a cold frame for the winter. Many will germinate the following spring, but some take another year to sprout. If sowing inside, stratify the seeds in their packet or an envelope for at least 6 weeks. Afterward, soak and plant them as above. Hold their containers at temperatures of 10° to 16°C. They will germinate erratically, some taking a year to do so.

Additional methods: *Side-veneer or side-wedge grafting.* Make the graft in early to mid-spring, and protect it from wind and passing animals to ensure that it stays in place while the graft is forming.

Hardwood cuttings. Take these cuttings in late winter. They require bottom heat while they are rooting.

Potential Problems

Seeds germinate erratically enough so that gardeners sometimes lose patience with them and throw out the tray prematurely. Resist this temptation – germination can take as long as a year. If grafting, remember to keep the air temperatures cool to retard the growth of the scion while the graft is forming. When starting cuttings, use only new, clean media to protect against rot.

Abutilon spp.

Flowering Maples
MALVACEAE

Zones: 8–10

The 150 species in this genus come from tropical and subtropical regions in North and South America, Asia, Australia and Africa. Some species are deciduous, while others are evergreen, and the genus includes shrubs, small trees, perennials and annuals. Most species have very showy, cup- or bell-shaped flowers in colours ranging from white to orange, yellow and pale red. Some species have prominent, vividly coloured calyces, while the leaves of others, such as *A. pictum* 'Thompsonii', are variegated with a pale yellow. Many hybrids such as 'Kentish Belle' and 'Ashford Red' are well-loved houseplants, particularly because they are reliable bloomers during dreary winter days.

PROPAGATION METHODS

Easiest: *Seed.* Start seeds indoors in early spring. Maintain soil temperatures of 21° to 24°C. Seeds should germinate in about 3 weeks.

Additional methods: *Semiripe cuttings.* Take these when they are ready, generally in midsummer.

Softwood cuttings. Take these as soon as the shoots are 10 to 15cm long and have not yet begun to harden.

Potential Problems

Seeds with a long germination period, such as flowering maples, are especially prone to fungal attack. Protect seeds by planting them in trays filled with a nutritious soil mix covered with 0.5 to 1cm of moistened vermiculite. Make the planting furrows in the vermiculite, and cover the seeds with vermiculite. Protect cuttings from fungal attack by starting them in a new, clean, soil-less medium, and use bottom heat to maintain soil temperatures of 24° to 27°C.

Acanthus spp.

Bear's Breeches
ACANTHACEAE

Zones: 5–9

The 30 species of striking perennials in this genus come from dry, rocky places, mostly in the Mediterranean. Their tall flower spikes, which can reach a height of 1.3 m, range in colour from white to pale green and yellow to

pink. Some species have pink or purple flower bracts, making the blooms appear bicoloured. The glossy and deeply lobed leaves can grow 1 to 1.3 m long and add majesty to the garden even when the plant isn't in flower.

PROPAGATION METHODS

Easiest: *Seed.* Start seeds inside in early spring. If starting *A. mollis* or *A. spinosus*, soak the seeds for 24 to 48 hours. After planting, keep the soil mix at a temperature of 4° to 7°C. Germination can be erratic, but most of the seeds should germinate within 3 months. When starting *A. hungaricus,* soak the seeds as above, but maintain soil temperatures of 18° to 21°C for 2 to 3 weeks, then lower the temperature to -4° to 4°C for a month to 6 weeks, and then raise the soil temperature again to 4° to 7°C.

Additional methods: *Division.* Divide plants in early spring, before they have had an opportunity to put on new growth.

Root cuttings. Take these cuttings in late autumn or early winter.

Potential Problems

Because the seed takes a long time to germinate, it can be prone to fungal attack. It's also possible to forget about it and let the tray go dry. Place it in a spot where it's easy to maintain the appropriate soil temperatures and you'll remember to monitor it carefully. Divisions can also dry out; keep divisions covered with damp hessian or newspapers if you do not immediately pot them up or transplant them. Root cuttings sometimes fail simply because their medium dries out over the winter or is so wet that fungi attack. Monitor the pots or trays where you've planted the root cuttings all through the winter and into the following season.

Acer spp.

Maples
ACERACEAE

Zones: 3–9

The 150 species in this genus come from Europe, North Africa, Asia and North and Central America. Maples are enormously varied. Most species are deciduous, but some are evergreen. During the growing season, leaves can be variegated and are light to dark green, bronze or the reddish colour of the Japanese maples (*A. japonicum, A. palmatum*). In autumn, most turn vivid shades of yellow, orange or red. Flowers are inconspicuous, but seeds are enclosed in winged fruit – "whirligigs" – that spiral through the air as they fall to the ground. Bark is as variable as the leaves. The bronze-coloured bark of paperback maple (*A. griseum*), for example, peels away in the same manner as that of birch trees, while striped maple (*A. pensylvanicum*) has smooth green-and-white-striped bark.

PROPAGATION METHODS

Easiest: *Seed.* Collect seeds as the winged fruits fall to the ground, and plant them immediately. Two species, *A. rubrum* and *A. saccharinum,* produce seed in spring. These will germinate easily, within 2 weeks. Give seeds of other species a head start by soaking them 24 to 48 hours before planting. Most species germinate well if you hold their planting medium at a temperature of 21°C for a month and then move them to a location where the soil temperature can maintain temperatures of 2° to 4°C. Germination usually takes 3 to 6 months.

Additional methods: *Whip-and-tongue grafting.* Graft in late winter.

Side-veneer grafting. Graft in late winter.

Softwood cuttings. Although this technique is not as surefire as starting from seed or grafting, you may have good luck propagating maples with softwood cuttings taken in mid-spring to early summer.

Potential Problems

Seeds with a long germination time can easily fall prey to fungal attack. Start maple seeds in deep containers—either cell packs designed for tree seeds or pots at least 10cm deep. Fill the containers three-quarters full with a nutrient- and humus-rich soil mix, and top that with a 2.5cm layer of vermiculite. The seeds will germinate in this area, reducing the chances of disease. Grafts are likely to take

well unless they dry out. Make certain to wrap them securely to prevent the cut surfaces from being exposed to air. Maples tend to transpire a great deal. When rooting softwood cuttings, remember to pull off the bottom leaves and cut the remaining leaves in half to reduce the amount of water they can lose.

Achillea spp.

Yarrows
ASTERACEAE

Zones: 3–8

The 85 species in this genus come from North America, Europe and other temperate regions. In warm climates, the plants are perennial, but they are grown as hardy annuals in colder areas. Most species are grown for their flower heads, which grow in flattened clusters at the top of stems. Individual florets are small and range in colour from white to yellow, orange, pink, red and bronze shades. Leaves of most species are aromatic and range in colour from dark green to grey and silver tones. If you're looking for yarrows that are especially long lasting when cut, grow *A.* x 'Coronation Gold', *A.* x 'Moonshine', or one of the Galaxy hybrids.

PROPAGATION METHODS

Easiest: *Seed.* Start seeds early indoors for blooms in mid- through late summer and early autumn. Plant in furrows filled with vermiculite, but don't cover – the seeds need light. The soil mix should be a steady 21°C while plants

are germinating. They will germinate in 1 to 2 weeks, at the most. Grow on at 10° to 16°C until time to transplant to the garden.

Additional methods: *Division.* Divide mature clumps in early spring, before shoots have begun to grow.

Softwood basal cuttings. Take softwood cuttings from shoots that grow at the base of the plant when they are about 15cm long. Root them in a soil-less medium. They root quickly.

Potential Problems

Seeds won't germinate if they don't see light. After planting, mist them lightly so they find niches in the vermiculite where they can remain moist but still see light. Cover the trays with plastic wrap to retain moisture. Divide clumps into no more than 2 or 3 pieces; smaller pieces may survive but will take a few years to become large enough to make a visual impact in the garden. Cuttings won't root properly if they are dry. Mist the cuttings frequently to keep the air around them humid.

Adiantum spp.

Maidenhair Ferns
PTERIDACEAE

Zones: 3–10

The majority of the more than 200 species in this genus come from tropical and subtropical regions of North and South America, although a few hardy species hail from colder areas in Europe, Asia and the North American

Achillea spp.

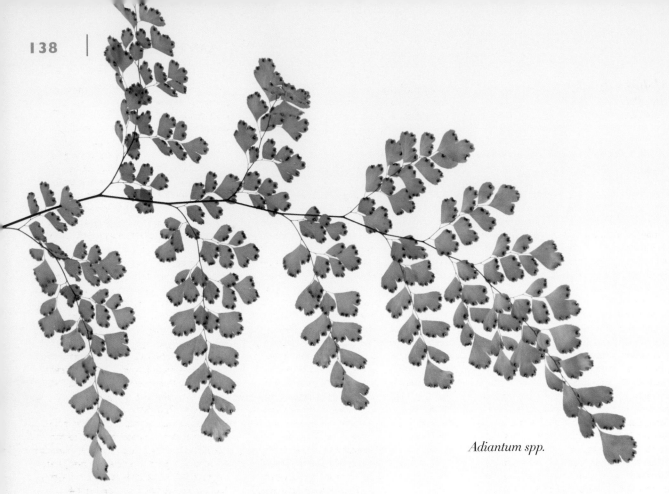

Adiantum spp.

continent. The tropical species are evergreen, but hardy species, such as the American maidenhair fern (*A. pedatum*), die back to the ground in late autumn. These ferns are grown for their drooping fronds with delicate, rounded segments. Fronds of some species are reddish when young, but most are a fresh, medium green colour.

PROPAGATION METHODS

Easiest: *Division.* Maidenhair ferns grow from rhizomes that are easy to divide. Dig and divide them in early spring, before new fronds are growing vigorously. Plant the divisions immediately to prevent their drying out.

Additional methods: *Plantlets.* Trailing maidenhair (*A. caudatum*) fern forms plantlets at the tips of its fronds. Place pots of humus-rich soil mix under the plantlets, and peg them in place. They will form roots within a month or so. After roots are well established and growing vigorously, sever plantlets from the parent plant. Transplant plantlets to the garden or a larger pot when environmental conditions are correct – not in the heat of summer.

Spores. It is possible to grow plants from spores, but environmental control has to be impeccable. Don't attempt it unless you have a reliable propagating chamber with mister.

Potential Problems

Divisions are just about foolproof as long as you remember to take a large enough piece of the rhizome so that the new plant will have both adequate roots and fronds.

Ajuga **spp.**
Bugles
LAMIACEAE

Zones: 3–8

The 40 species in this genus come from temperate areas in Europe and Asia. A few species are annual, but most are perennials. They thrive in somewhat shady spots and are widely used as ground cover under a dense canopy of tree leaves or in shade. Carpet bugle (*A. reptans*) is the most commonly grown and includes many cultivars. The species itself has dark green leaves and spikes of dark blue flowers in late spring and early summer, but 'Burgundy Glow', 'Catlin's Giant', and 'Multicolour' all have red, bronze or purplish colouration on their leaves.

PROPAGATION METHODS

Easiest: *Division.* Species such as *A. reptans* grow long stolons. These stolons take root where they touch the ground and also send up new shoots, so they propagate themselves. Simply cut off a rooted plant and transplant it to a new location. Other species, such as *A. pyramidalis* and *A. genevensis*, don't form stolons, so you'll need to divide them by cutting through their rhizomes.

Additional methods: *Seed.* If you are interested in collecting seed, you'll need to bag the flower stalk before the petals fall. Use cotton at the base of the bag to catch any falling seeds, and let the seeds naturally dry on the plant. If you live in a cold-winter area, stratify the seeds for at least 6 weeks before planting in early spring. Seeds require light to germinate and do best at a soil temperature of 13° to 18°C.

Potential Problems

The species that grow from stolons are foolproof – if you wait for the plantlets to develop a root system large enough to support the top growth before cutting it from the mother plant. Secure the bottom of the bag you place over the seeds and make sure that the cotton ball is in place to catch them.

Alchemilla spp.
Lady's Mantles
ROSACEAE

Zones: 3–8

The roughly 250 species in this genus come from northern temperate areas in Greenland, as well as mountainous areas of northern Europe, Africa, India, Sri Lanka and Indonesia. They are grown both for their deeply lobed and decorative leaves as well as for the clusters of small, greenish yellow flowers that rise above the leaves each spring. The flowers last for a week when cut, and the plants are excellent in rock gardens and make good ground cover in well-drained soils. 'Thriller' is a cultivar that's prized for its star-shaped flowers with a strawlike texture; it's often grown as a cut flower.

PROPAGATION METHODS
Easiest: *Self-seeding*. Plants self-seed so easily that most gardeners don't have to do more than dig up and transplant the volunteers they discover. In fact, you may come to curse this plant's easy productivity.

You can purchase seeds for some cultivars. Plant them early inside. They require light to germinate and respond best to daytime soil temperatures of 18°C and nighttime temperatures of 10° to 13°C. They will germinate in 1 to 3 months, so prick out seedlings as they germinate, but don't throw out the tray.

Additional methods: *Division*. Divide plants in early spring, before they have had an opportunity to send up flower stalks. Snip off the flower stalks after transplanting the divisions so the plant can put its energy into becoming established.

Potential Problems
Lady's Mantle is susceptible to fungal diseases if planted in soggy soils. Make sure to place your divisions in well-drained areas to avoid these problems. Seeds do not germinate well without alternating day and night temperatures. Take the time to move them in and out of a propagating unit or on and off a heating mat – you'll be rewarded for this extra work.

Alisma spp.
Water Plantains
ALISMATACEAE

Zones: 5–8

The nine species in this genus come from temperate areas in the northern hemisphere, southern Africa and Australia. They thrive in boggy spots and at the edges of ponds and swamps. In water gardens, they are grown for their striking, lance-shaped foliage that's held above the water by long, purplish brown leaf stalks. In late summer and early autumn, tiny white or slightly pink flowers rise above the foliage, giving the plant an airy appearance.

PROPAGATION METHODS
Easiest: *Division*. Plants grow from rhizomes, so they are easy to divide. Take divisions in mid-spring to early summer.

Additional methods: *Seed*. You can purchase seeds from some suppliers of aquatic plants, but it's easy to collect your own from a blooming plant. Bag a group of flowers just as the petals are beginning to fall. Use cotton balls at the bottom of the bag, and tie it off tightly so you don't lose falling seeds. Plant the seeds immediately in a rich soil mix covered with a slick of water. They require darkness and high soil and water temperatures to germinate. Expect to see new plants within 2 weeks to a month.

Potential Problems
Never let these plants dry out. When dividing, keep their rhizomes underwater, and move them quickly to their new sites. Don't let the seeds dry either – they are built to live in water.

Allium spp.
Onions, Ornamental Onions
LILIACEAE

Zones: 3–10

The more than 700 species in this genus come from mountainous areas in the northern hemisphere. Many, such as chives (A. schoenoprasum) and garlic chives (A. tuberosum), are edible, but most are grown for their striking flower heads. Leaves are tall, generally hollow spikes that give off an oniony odour when crushed. Flower heads range from vividly coloured ball-shaped blooms, as in A. caeruleum and A. giganteum, to more loosely formed clusters, such as those of A. cristophii

and A. flavum. Various species bloom from spring to late summer.

PROPAGATION METHODS
Easiest: *Division*. Some species produce rhizomes that can be cut into pieces and replanted in early spring, while others produce bulblets around the mother bulb in late summer and autumn. Plant the bulblets in nursery beds, and mulch them well over the winter. In spring, remove the mulch to let the soil warm, and keep the nursery bed moist. They will germinate within a few weeks to a year, so don't disturb the area.

Additional methods: *Seed*. Seed is available for many species of this genus. It requires a 2-week chilling period at about 4°C. Mix seeds with moist vermiculite, and let them sit in the refrigerator for this time. Barely cover the seeds with vermiculite when you plant them – they need light to germinate. Soil temperatures of 10°C are ideal. Most germinate within a few weeks, although you should leave the trays in place for a couple of months to account for stragglers.

Potential Problems
Plant baby bulbs at a depth of only two to three times their diameter to make certain that their emerging leaves can grow above the soil surface. They are resistant to many pests but extremely susceptible to various fungal diseases. Protect them by planting only in well-drained soils. Seeds do not germinate well at high soil temperatures. Plant early enough in the season so that they will germinate while it's still easy to keep their soil temperatures cool.

Aloe spp.
Aloes
ASPHODELACEAE

Zones: 9–11

The 300 or more species in this genus come from tropical and subtropical areas of the Cape Verde Islands, Africa, Madagascar and the Arabian Peninsula. A. vera, or medicinal aloe, is commonly grown as a houseplant in northern areas because the gel in its succulent leaves has a healing effect on burned skin. It is said to confer to the body many benefits. Some species, such as A. haworthioides, A. variegata and A. aristata, are grown primarily for their decorative foliage, while others, such as A. arborescens and A. mitriformis, have blooms that equal the foliage in beauty.

PROPAGATION METHODS

Easiest: *Offsets.* Most aloe plants produce offsets around the mother plant at the end of the rainy season. Separate these when they can be removed with adequate roots – roughly as long as the top growth – to survive on their own. Repot them in cactus mix. If growing them in a zone where they are hardy, transplant them to the garden at the beginning of the next rainy period. Otherwise, keep them well watered while they are becoming established in their pots, and then treat them as you do their parent stock.

Additional methods: *Leaf cuttings.* Most aloe plants propagate well from vertical leaf cuttings. Start them in a fast-draining, soil-less medium, and keep them out of direct light while they are rooting. Once they have formed roots, transplant them into a cactus mix and move them into an area with high light levels.

Seed. Seeds are available for some species. They require light to germinate, so mist them into niches in a vermiculite covering over a tray filled with cactus mix. Keep their flats at a soil temperature of 21° to 24°C.

Potential Problems

Soggy soil kills aloes. Make certain to plant offsets only in fast-draining mixes. However, remember that in nature, most of these plants live in areas with an annual rainy season – generally in the summer, when temperatures are high. So don't neglect watering them; just make sure that the mix drains rapidly enough so that roots are not exposed to prolonged periods of wet soil.

Amelanchier spp.

Juneberries, Serviceberries
ROSACEAE

Zones: 3–9

The 25 species in this genus come from moist wooded areas in Europe, Asia and North America. They are primarily grown as ornamental shrubs because of their naturally graceful habit, lovely spring flowers and brightly coloured autumn foliage. However, don't neglect their berries – they have a bright taste and, if they are too tart for you when eaten raw, make excellent preserves or additions to pies or muesli. *A.* x *grandiflora* is one of the most popular species and boasts numerous cultivars, including 'Princess Diana', 'Autumn Brilliance', and 'Robin Hill'. Running

serviceberry, or *A. stolonifera*, has particularly sweet fruit, as does *A. lamarckii*.

PROPAGATION METHODS

Easiest: *Suckers.* Most Juneberries form suckers that can be severed from the plant when they are a year or so old. Dig down and expose the base of the sucker to make certain that it has an adequate root system before severing it from the parent plant.

Additional methods: *Layering.* Juneberries form roots from the nodes on their branches so readily that you can layer them with ease. Pin a branch to the ground in early summer, and leave it in place until the following spring. By then, it will have an extensive root system and be ready to grow on its own.

Root cuttings. If you want to propagate this plant in winter, take root cuttings. Plant them in a well-drained soil mix, and leave the pots or trays outside over the winter months. New stems and leaves will form in late spring to early summer of the following year.

Potential Problems

These plants are close to foolproof, especially if you are propagating them from suckers or layering them. The only possible problems you might have would be those caused by soggy or nutrient-poor soils. Make certain that the soil drains well and contains moderate levels of humus-rich compost.

Anemone japonica spp.

Wind Flowers
RANUNCULACEAE

Zones: 6–10

The 120 species in this genus come from temperate areas in both the northern and southern hemispheres. All are perennials and characterized by open, composite flowers that have a central disk surrounded by rays of white, yellow, pink, red, purple and blue. The flowers, which grow on long stems that hold them well above the foliage, dance in spring or late summer breezes. *A. blanda* is a favourite species in temperate gardens and includes many popular cultivars, including 'Atrocaerulea', 'Charmer', 'Violet Star', and 'White Splendor'. *A. coronaria* 'The Bride' is a favorite in Zones 8 to 10, and cultivars in the De Caen group are often grown in northern areas for their vividly coloured single flowers.

PROPAGATION METHODS

Easiest: *Division.* Plants may grow from tubers, rhizomes or fleshy roots. No matter which, they are easy to divide. If they bloom in spring, wait until they have gone dormant for

Anemone spp.

the year to lift and divide them; if they bloom in autumn, divide them in early spring, and hold them in a sheltered spot in containers until the following year. Prevent them from blooming during this time – you'll want them to put their energy into forming roots, not flowers.

Additional methods: Seed. Plant seeds as soon as they are ripe in trays, and hold the trays in cold frames over the winter. If you cannot collect your own seed, buy from a good supplier – they send seed in foil packets in autumn. After the seed has overwintered in the cold frame, bring it indoors or place it in a sheltered position where you can keep it moist and at a temperature of 18° to 21°C. It should germinate within a month. If it does not, leave it for another year.

Root cuttings. Autumn-flowering species propagate well from root cuttings taken in early spring. They generally produce stems and leaves by late summer. Hold them in a cold frame over the winter, and plant them out the following spring.

Potential Problems

Protect new divisions from fungal diseases by planting only in well-drained soils. They require moderate levels of humus, too, and thrive when compost or leaf mould is added to their planting holes. Some seeds are covered with fine hairs. They will germinate better if you rub off the hairs with your fingers. Don't use sandpaper because they die if the seedcoat is damaged.

Artemisia spp.
Wormwoods, Mugworts
ASTERACEAE

Zones: 3–9

Most of the 300 or so species in this genus come from the northern hemisphere, but a few are native to South Africa and South America. They are grown for the ornamental value of their aromatic leaves, which are green to grey-green and silver in colour and finely cut to deeply lobed to almost threadlike in shape. *A. absinthium* was used to make the drink absinthe, southernwood (*A. abrotanum*) is used in sachets to repel clothes moths, sweet Annie (*A. annua*) is an annual form used to make dried wreaths, 'Powis Castle' is used as a mounded accent in beds and borders, and tarragon (*A. dracunculus*) is a culinary herb.

PROPAGATION METHODS

Easiest: *Division.* Divide perennial plants in early spring. Their roots are tough – you'll need to use a sharp garden knife to make

clean cuts through the root mass.

Additional methods: Semiripe cuttings. Take cuttings in mid- to late summer, and root them in a soil-less mix out of direct sunlight. Grow them inside or in a greenhouse until the following spring.

Seed. Annual wormwoods can be started from seed. Plant them in early spring. They require light to germinate and soil temperatures of 16° to 18°C.

Potential Problems

Most wormwoods grow well in dry, rocky conditions. As a consequence, they cannot tolerate soggy soils. However, when you plant divisions, place them in well-drained areas, but make certain that the young plants have adequate moisture while they are getting established. Do not let sweet Annie go to seed in the garden; it self-seeds so vigorously that you'll be pulling it out of every area of your garden – for years! Your neighbours will be equally well supplied with this plant; not all of them will be delighted.

Asarum spp.
Wild Gingers
ARISTOLOCHIACEAE

Zones: 4–9

The 70 species in this genus come from Europe, Asia and North America. They are named wild ginger because their rhizomes smell a bit like ginger, but they are not edible. They are used as ground cover in shady areas because the leaves are large, often shiny, and sometimes have pale veins that emphasize their beauty. Blooms are hidden under the leaves and have a somewhat unpleasant odour. Canadian wild ginger (*A. canadense*) is deciduous, while European wild ginger (*A. europaeum*), *A. hartwegii*, and *A. shuttleworthii* are evergreen. 'Callaway' is a popular cultivar of *A. shuttleworthii* because of its highly decorative leaves and tolerance of shady conditions.

PROPAGATION METHODS

Easiest: *Division.* Divide plants in early to mid-spring. Be gentle when digging and dividing these plants because their rhizomes can be brittle.

Additional methods: *Seed.* Seed must be planted as soon as it is ripe. Buy only from a supplier who can promise to send the seeds in mid- to late summer, just after they ripen. Plant the seeds in trays, and place them where you can maintain soil temperatures of 16° to 18°C. They will germinate in 1 to 3 weeks.

Grow them in a protected spot over the winter, and transplant them to their permanent location the following spring.

Potential Problems

Plant divisions as soon as you make them – don't let them dry out. Wild ginger likes moist, rich soil with high humus levels and is intolerant of dry conditions. Old seed will not germinate. If you have trouble finding a good seed supplier, search for someone in your neighbourhood who has a plant you admire and ask if you can save some of the seeds. Bag the flowers when they begin to wilt but before the petals fall. Watch the seeds carefully so you can plant them just after they fall out of the seed pods.

Asparagus spp.
Asparagus Ferns
LILIACEAE

Zones: 4–11

The 300 species in this genus come from Europe, Asia and Africa. Most species are grown for their ornamental value, although *A. officinalis* is the vegetable that graces spring tables all over the world. In northern areas of the United States, ornamental asparagus ferns such as *A. setaceus*, *A. densiflorus* 'Sprengeri' and *A. densiflorus* are commonly grown in green-houses so that cut foliage can be added as a filler in bouquets of flowers. These plants are often grown as houseplants, too, most commonly in hanging baskets. In Zones 9 and 10, these plants are common weeds that grow in waste places, along fence lines, and wherever else birds – who love the berries – drop them.

PROPAGATION METHODS

Easiest: *Division.* Plants grow from underground tubers and are easy to divide. Take divisions just before spring if growing as a houseplant and just before the rainy season if growing outdoors.

Additional methods: *Seed.* Seeds are available for most types of asparagus ferns. Soak the seeds for at least 24 hours, and then rub them lightly on sandpaper to scarify them. Plant them in a moist, humus-rich mix, and set them where you can maintain soil temperatures of about 16°C, both night and day. They will germinate in 3 weeks to a month.

If you want to grow *A. officinalis*, start it from seed as described above, and let it develop in a nursery bed for 2 years. Once it blooms, you will be able to identify the male plants – the ones without berries. Transplant these to your perennial food garden because they will be more productive than female plants over the coming years.

Potential Problems

These plants grow so easily that it's no wonder that they are a weed in areas where they can grow wild. You are unlikely to have problems with them unless you let divisions or seedlings dry out while they are becoming established. Once they are growing well, they can withstand a certain amount of drought.

Asperula spp.
Woodruffs
RUBIACEAE

Zones: 5–9

The 100 species in this genus come from Europe and Asia, generally from mountainous areas with thin soils and good drainage. The genus includes annuals as well as evergreen and deciduous perennials and ranges in habit from mat-forming ground cover to small, tidy shrubs. *A. suberosa* and *A. orientalis* both have much showier flowers and also tolerate filtered shade well.

PROPAGATION METHODS

Easiest: *Seed.* Seeds are widely available for several species of woodruff. Cool the seeds in the refrigerator for at least 2 weeks before planting them. They can take up to a month to germinate, so it's wise to start them at least 8 weeks before the frost-free date. Plant them in a quickly draining soil mix covered with a thin dusting of fine vermiculite. Sprinkle them on the soil surface, and then mist them so they fall into moist crevices in the vermiculite. They require light to germinate, so cover trays with plastic film. Place them where soil will maintain temperatures of 10° to 16°C.

Additional methods: *Division.* In early spring, lift and divide established plants. Basal cuttings are also effective if you want to propagate plants without disturbing them. When rooting cuttings, strip the leaves from the bottom 2.5cm of the stem. Use rooting hormone, and place the cuttings in fresh soil-less rooting medium. Place in indirect light. They should root within a month or so.

Potential Problems

Plants are so easy to start from seed that you're unlikely to have any trouble with them. They often self-seed, not only in their own area of the garden, but also where animals have dropped seeds that clung to them as they passed through the woodruff patch. Woodruff tolerates thin soils, so people often forget to water new transplants. However, until they are established, divisions need consistently moist soil. Keep air circulation high around rooting cuttings to avoid problems with fungal diseases.

Astilbe spp.
Astilbes
SAXIFRAGACEAE

Zones: 4–8

The 12 species in this genus come from southeast Asia and North America. Although there are only a few common species, the number of hybrids is breathtaking. Many of the most common are hybrids of *A.* x *arendsii*, although you're likely to see hybrids of *A. chinensis, A. japonica, A. simplicifolia* and *A. thunbergii* as well. Plants are grown for their airy plumes of flowers that wave over decorative leaves in late summer. Flower colours range from white through yellow, coral, apricot, pink, red and magenta, and leaves sometimes have red veins and petioles.

PROPAGATION METHODS

Easiest: *Division.* Divide in early to mid-spring. The new growth of astilbe is slow to appear in early spring, so you have plenty of time to divide them before they resume new growth.

Additional methods: *Seed.* Seeds are available for some cultivars. Although they germinate easily, they are so tiny and grow so slowly that you should start them from seed only if you are doing so to get a particular cultivar that is otherwise unavailable or you love horticultural challenges. Sprinkle seeds on the top of a thin layer of vermiculite that covers a fine-textured potting mix. Mist them in place, and cover the tray with a pane of glass, not plastic – you want it to be elevated above the surface of the medium. Place the flat in indirect light and over a heating mat set at a steady 18°C. They can take anywhere from 30 to 90 days to germinate. When you finally do see the tiny little leaves, pull the glass back a bit to allow some fresh air into the tray, but do not entirely remove it. Gradually pull it back over the next few days. After that, mist the tray with the finest sprayer you have. Plants will take a few months to grow large enough to plant in 7.5cm pots. Don't transplant to the garden until late summer or early autumn, and mulch them well for the winter months.

Potential Problems

Plants are somewhat fragile, so divisions must be handled with care. Keep them moist while they are becoming established, and make certain that their planting holes contain lots of humus. Leaf mould is a good soil amendment for these plants.

B

Baptisia spp.
Wild Indigos, False Indigos
FABACEAE

Zones: 4–8

The 20 species in this genus come from North America, specifically in the area of the United States. Their common name comes from the striking blue colour of the blooms of *B. australis* – a clear indigo. Other species are likely to have white or cream blooms, but all bear their flowers on tall spikes that rise above the leaves. Both *B. pendula* and *B. lactea* have grey-green leaves and white flowers that are followed by beautiful hanging seed pods that

Begonia spp.

ripen to a rich black. *B. perfoliata* has yellow blooms and grey-green leaves.

PROPAGATION METHODS

Easiest: *Seed.* Begin by rubbing the hard seedcoat against fine sandpaper to scarify it, and then soak it in warm water for 6 to 8 hours. Plant it in a water-retentive and nutritious soil mix, covering it with about 0.5cm of soil. Place the tray on a heating mat to maintain steady temperatures of about 21° to 24°C. Germination will take place in about a week.

Additional methods: *Division.* Lift and divide plants in early spring.

Potential Problems

When scarifying seeds, be careful to injure only the seedcoat, not the interior tissues of the seed itself. Keep germinated seedlings in a position with good air circulation because they are susceptible to damping off and other rotting diseases. Immediately plant divisions. Do not let them dry because the plant must have adequate moisture while it is becoming established. Once it is well rooted, it can tolerate dry soils.

Begonia spp.
Begonias
BEGONIACEAE

The more than 1,300 species in this genus come from tropical and subtropical areas of Africa, South and Central America and Australia. While the number of species is enormous, the number of hybrids is even more astounding. Begonias are generally broken into seven groups: cane-stemmed, rex-cultorum, rhizomatous, semperflorens, shrublike, tuberous and winter-flowering. The leaves of all species are decorative – some large, pointed and marbled in colouration; and others small, rounded and waxy green to a deep reddish colour.

PROPAGATION METHODS

Easiest: *Cuttings.* This is the easiest way to propagate all types of begonias. Greenwood cuttings taken just before spring root quickly in a fast-draining medium or, in the case of cane-stemmed cultivars, even a glass of water. Rex begonias are known for their ability to grow easily from leaf-vein cuttings (see page 90).

Additional methods: *Division.* Begonias that grow from rhizomes or tubers are easy to

divide. Separate tubers when you repot plants in fall and divide rhizomes in early spring, again as you repot the plants.

Seed. Seeds for semperflorens and rex begonias are widely available. Start them in early spring in a humus- or peat-filled soil mix covered with a thin layer of vermiculite. They require light to germinate and prefer soil temperatures of 21° to 27°C.

Potential Problems

Seeds grow slowly and are susceptible to various fungal diseases, so it's important to grow them in an area with high levels of air circulation. Cuttings may also rot if their medium is kept too wet and the air is stagnant. Root them only in fresh, soil-less mixes, and grow them in a well-lit area out of direct sunlight.

Belamcanda spp.
Blackberry Lilies
IRIDACEAE

The two species in this genus come from India, China, Japan and other mountainous regions in the Far East. Although members of the Iris family, they look more like lilies, as their common name suggests. They are grown for their long, strap-shaped leaves and branching stems with showy flowers. The yellow and orange flowers are marked with dark maroon spots. The seed pods open to reveal shiny black seeds – the "blackberries" of their name.

PROPAGATION METHODS

Easiest: *Seed.* Plant seeds in a compost-rich soil mix as soon as they fall from the pods. Set their planting trays in a cold frame over winter. Keep them moist, and prevent them from overheating. They will germinate in spring. Alternatively, collect the ripe seed and place it in a plastic bag in the refrigerator over the winter. Plant it in early spring, and give the tray day temperatures of 24° to 29°C and night temperatures of 16° to 18°C. Seeds will germinate in about a week.

Additional methods: *Division.* Plants grow from rhizomes that can be cut apart and divided into smaller sections. Because blackberry lilies tend to be short-lived, it's wise to divide the rhizomes every 2 to 3 years, discarding the older, woodier rhizomes.

Potential Problems

If planting the seeds in trays and overwintering them outside, it's easy to forget about them and let the cold frame overheat. To prevent

Bergenia spp.

doing this, place the tray in a protected spot under a porch or against a north wall and cover it with a pane of glass and then a 30cm-deep layer of mulch for the winter months. Remove the mulch in early spring, before the frost-free date but after the worst of the spring storms have passed. Don't let divisions dry out when you are working with them. Choose a cloudy day to dig them, and keep unplanted rhizomes covered with damp newspapers or hessian.

Berberis spp.
Barberries
BERBERIDACEAE

Zones: 3–9

The 450 species in this genus come from the northern hemisphere, Africa and South America. All species are woody shrubs, but some are deciduous and some are evergreen. Their leaf forms are varied; some have toothed leaves with thorns at the margins, and some have shiny, smooth, oval leaves. They range in colour from a soft green to olive green, bronze, wine and variegated rose and burgundy. Most have spectacularly colourful autumn leaves in shades of bright red and yellow. Bloom colours also vary, but most species have yellow flowers. In autumn, pink, red, blue or black berries provide winter food for birds. This alone is reason enough to grow barberries.

PROPAGATION METHODS
Easiest: *Semiripe cuttings.* Take cuttings of deciduous and evergreen barberries in mid- to late summer, as soon as they become semi-ripe. Root them in a clean, soil-less medium. Softwood cuttings, taken in summer, are a reliable method for propagating deciduous cultivars.

Additional methods: *Mallet cuttings.* Take mallet cuttings in summer, and root them in fresh soil-less mix.

Seed. Seeds are available for some cultivars, and you can also save your own seed. If more than one barberry species grows in the area, you're likely to get a hybrid plant from self-saved seed. But that's not all bad – you might come up with a plant with especially desirable characteristics. Plant seeds in outdoor nursery beds. Cover them lightly, mark the rows well, and keep them well watered. They may germinate erratically.

Potential Problems
Fungal infections are always a threat to cuttings. Make certain to use an initially sterile starting mix, and monitor it closely. Remove cuttings that look diseased to protect others.

Bergenia spp.
Elephant's Ears, Pigsqueaks
SAXIFRAGACEAE

Zones: 3–8

The eight species in this genus come from central and eastern Asia. They are widely grown for their large, leathery leaves that remain throughout winter in most climates and their clusters of spring-blooming flowers. These plants tolerate filtered shade well and grow well on the edges of wooded areas. Some cultivars, such as 'Ballawley', have wine-coloured leaves, while others, such as 'Bressingham Bountiful', have dark green leaves with maroon edging. 'Bressingham Ruby' is known for its green leaves with maroon undersides that turn bright red in autumn and winter. Blooms range in colour from white to pink to dark red, depending on the cultivar.

PROPAGATION METHODS
Easiest: *Division.* These plants grow from thick rhizomes that are easy to divide just after they flower or in early autumn.

Additional methods: *Seed.* If you save seed from these plants and grow more than one cultivar, you can expect to produce new hybrids. Experiment with saving seed and growing it out – you may be very pleasantly surprised by the results. Seeds require stratification, so it's best to hold them over the winter in containers in the refrigerator. Plant them in a mix containing peat moss or compost. Place them on a heating mat set for 18°C. They should germinate in 1 to 3 months.

Potential Problems
The rhizomes decline with age, so it's wise to divide them every 3 to 5 years. Don't divide them until after they bloom, though; the plant won't have the resources to grow new roots until then. Seeds start quite easily and are very attractive to slugs. Don't plant out new plants without giving them some protection from any slugs or snails in the neighbourhood.

Buddleja spp.
Butterfly Bushes
LOGANIACEAE

Zones: 5–9

The nearly 100 species in this genus come from Asia, Africa and North and South America. The genus includes shrubs, climbers and a few herbaceous perennials; some are evergreen, but most are deciduous. Butterfly bushes are grown for their blooms, which, true to their name, attract butterflies from miles around. The tiny florets of most species form long cones, as does the well-loved *B. davidii*, but in others, such as *B. alternifolia*, the florets make clusters grouped in panicles on a long stem, and still others – *B. globosas* for example – grow in globes on individual stems. Flowers range in colour from blue to white, pink, purple and yellow.

PROPAGATION METHODS

Easiest: *Seed.* Seeds of many cultivars are widely available. Stratify seeds for at least 2 weeks in the refrigerator. Plant inside in very early spring, and place the starting containers on a heating mat that keeps the soil temperature around 16° to 24°C. They should germinate within 2 to 3 weeks.

Additional methods: *Semiripe cuttings.* Take semiripe cuttings in midsummer, as soon as they are ready to be cut.

Hardwood cuttings. These are ready in autumn. Plant them in trays, and hold them over winter in a cold frame or protected garden spot.

Potential Problems

Seedlings are slow growing. Protect them from fungal diseases by planting them in a fast-draining soil mix and keeping air circulation around them high. Transplant them to the garden only after they are 15 to 20cm tall; younger than that, they are too vulnerable to changes in the weather and animal damage. Protect cuttings from fungal diseases by rooting them only in fresh, sterile media.

Buxus **spp.**
Boxwoods
BUXACEAE

Zones: 6–9

The 70 species in this genus come from Europe, Asia, Africa and Central America. They are well-loved hedging shrubs because they are dense, evergreen and tolerant of pruning. These qualities make them ideal topiary subjects, and you often see them clipped into geometric shapes or fanciful animals in botanical gardens and estate plantings. Leaves of most species are smooth, oval and dark green, but cultivars such as 'Marginata' and 'Latifolia Maculata' are green with a yellow border, while the leaves of

'Elegantissima' are bordered with white.

PROPAGATION METHODS

Easiest: *Semiripe cuttings.* Take these cuttings in late spring or early summer, as soon as they have begun to develop a little woodiness at the base.

Additional methods: *Division.* Many boxes can be divided. Do this in spring, before the plant begins vigorous growth, and replant the divisions immediately.

Seed. Seeds are available for some cultivars, and you can also collect them from plants you admire. Remember that any seeds you collect are likely to be seeds of hybrids, so they won't be identical to their parent plant. However, you might get an interesting specimen this way. No matter if you collect or buy them, plant seeds in a cold frame in autumn. They will germinate sporadically the following spring and early summer. Grow them in containers until the following autumn, when you can transplant them to their permanent positions, or, if they are still quite small, hold them for another winter in the cold frame.

Potential Problems

Protect cuttings from fungal diseases by rooting them in a fresh, soil-less medium. Do not cut divisions too small, even though these are tough plants, they grow slowly enough so that they do not reestablish themselves quickly. Keep them well watered for the entire season after dividing.

Calla palustris
Calla, Bog Arum
ARACEAE

Zones: 4–8

The only species in this genus comes from temperate regions in the northern hemisphere. Callas can be deciduous or evergreen, depending on location. This species thrives in soggy soils on the edges of streams as well as in bogs and swamps. It has shiny green leaves and dramatic white spathes that surround the flowers. Red berries appear after the flowers.

PROPAGATION METHODS

Easiest: *Division.* Callas naturally reproduce by growing creeping rhizomes. If you want to transplant a section to a different site, you can simply cut off a section of the rhizome in early spring.

Additional methods: *Seed.* Aquatic suppliers sometimes sell calla seed. In late summer, as soon as the seeds are ripe, plant them in a nutrient-rich, humus-filled mix, and place the starting containers submerged to the rims in fresh water. They will germinate erratically, so don't remove the starting containers for at least a year.

Potential Problems

If divisions are too small, they will have trouble re-establishing themselves. Wait to divide them until the rhizome has begun to elongate naturally. Seeds may take some time to germinate; be patient, and don't give in to the temptation to throw out the pot.

Buxus spp.

Calluna vulgaris
Heather, Scotch Heather
ERICACEAE

Zones: 5–7

The one species in this genus comes from Europe, Turkey, Siberia, Morocco and the Azores. It is grown for its grey-green foliage and dense spikes of small, bell-shaped flowers that coat the plant. Blooms are red, pink, purple and white. All heathers make good ground cover but are especially attractive to bees so should not be planted in areas where children are likely to go barefoot. There are hundreds of cultivars. No matter what kind of heather you're looking for, it's likely that a speciality nursery will carry it.

PROPAGATION METHODS
Easiest: *Layering*. Plants layer naturally, so you simply have to help them along by pinning your selected branches to the ground in the spring. Check them in autumn; they may be ready to transplant to a new location, or you may want to wait until the following spring to sever them from the parent plant and move them.
Additional methods: *Semiripe cuttings*. Take cuttings in early to midsummer, as soon as they are ready. They will root quite quickly and should be ready to plant out by autumn.

Potential Problems
Wait to sever layered plants from the parent stock until they have a strong root system. Otherwise, they may not be able to acquire enough moisture to survive.

Calycanthus spp.
Sweetshrubs, Carolina Allspices, California Allspices
CALYCANTHACEAE

Zones: 5–9

The three species in this genus come from North America and grow wild along the margins of woods and on stream banks. They are grown for their compact habit and unusual flowers. The blooms are a dark dusky red or mahogany brown and grow at the ends of branches. They last for many weeks, spreading a fragrance similar to cloves through the garden. The leaves smell almost like camphor when crushed.

PROPAGATION METHODS
Easiest: *Suckers*. Suckers grow naturally. Dig them in early spring, and transplant immediately.
Additional methods: *Layering*. Plants also layer naturally, so it's easy to propagate them in this way. Bend the stems down to the ground in early spring, wound the bark and pin them in place. By the following spring, the new plants will be large enough to transplant easily.
  Softwood cuttings. Take cuttings in spring. They root quickly, but you may want to hold them in pots in a cold frame for one winter.
  Seed. Seeds are available from speciality companies. Plant them in autumn, as soon as they are ripe. Keep the seed trays in a cold frame over winter, and they will germinate in spring.

Potential Problems
Let both suckers and layered plants grow a strong root system before severing them from the parent plant. Otherwise, they may not be able to survive on their own. Seeds can germinate erratically; don't throw out the seeding tray.

Campanula spp.
Bellflowers
CAMPANULACEAE

Zones: 3–9

The more than 300 species in this genus come from temperate regions in the northern hemisphere, particularly Turkey and southern Europe. In addition to the huge number of species, there are hundreds and hundreds of cultivars. The genus includes annuals, biennials and perennials, some evergreen and some deciduous. They vary tremendously in habit; you can find trailing plants; spreading, mat-forming species; mounded forms; and tall, upright specimens. Flowers range in colour from white to pink, blue and purple; most are bell shaped, as are *C. persicifolia* 'Telham Beauty' and *C.* 'Burghaltii', but some (as in *C. carpatica* species) have a more open form.

PROPAGATION METHODS
Easiest: *Seed*. All species of *Campanula* are easy to grow from seed. Plant them on the surface of a vermiculite-covered tray, and mist them into small crevices. They need light to germinate. Cover the tray with plastic film until they germinate, which takes anywhere from 2 weeks to a month, depending on the species. Maintain soil temperatures about 16° to 18°C.

Additional methods: *Division*. All perennial species can be divided. Lift and divide them in spring or autumn, depending on where you live. Allow them at least 40 days to become established before the full heat of summer or the ground freezes in autumn.
  Softwood cuttings. Take softwood cuttings from shoots that grow at the base of the plant when they are about 15cm long. Root them in a soil-less medium. They root quickly.

Potential Problems
Seeds grow slowly, so they are prone to fungal diseases such as damping off. Keep air circulation high in the area where they are growing, and don't mist their leaves after about 2:00 p.m. – the leaves must be dry by nightfall. Divide plants so that each section has a portion of the root mass as well as top growth, and replant the divisions immediately. Keep them moist until they become re-established.

Campsis spp.
Trumpet Vines
BIGNONIACEAE

Zones: 5–9

The two species in this genus come from China and North America. This vine is grown for its bright orange or red trumpet-shaped blooms. The vines have a long bloom period, generally from mid- to late summer until early autumn, and are covered with flowers when in full bloom. In addition to its decorative qualities, this vine is grown because it is a favourite food of hummingbirds. Many people trellis it just outside a window so they can watch the parade of hummingbirds that visit every day.

PROPAGATION METHODS
Easiest: *Semiripe cuttings*. Take cuttings in early summer, as soon as they are ready. They root quickly and, unless you live in a climate with harsh winters, will be ready to transplant to the garden in autumn.
Additional methods: *Seed*. Seeds are available from speciality seed companies, and you can also save your own. Plant the seeds in a tray, and place it in a cold frame over the winter. Seeds will germinate erratically the following spring. Alternatively, you can stratify the seeds for 2 to 3 months in the refrigerator, and then start them inside the house. Place them on a heating mat set for 21° to 24°C. They will germinate in 4 to 12 weeks.

Hardwood cuttings. Take these cuttings in late autumn, as soon as the wood is truly hard. Hold them over the winter in a cold frame. They will root the following spring.

Root cuttings. Take cuttings just before the ground freezes in winter, and bring them inside the house or greenhouse to root. They should form roots in early to mid-spring.

Potential Problems

All cuttings are prone to fungal attack. Protect your plants by rooting them on a heating mat set at 16° to 18°C, using only fresh rooting media and keeping air circulation high in the area where they are rooting. Seeds will not germinate well if they do not receive a long chilling period. If you live where winters are mild, stratify them in the refrigerator rather than by having them spend the winter months in a cold frame.

Canna spp.

Cannas, Indian Shots
CANNACEAE

Zones: 8–11

The 50 species in this genus come from tropical and subtropical areas in Asia and North and South America. Cannas are spectacular specimen plants, thanks to both their dramatic leaves and striking flowers. Leaves can be as long as 61cm in cultivars such as 'Pretoria', marked with stripes of yellow or red. Flowers range in colour from yellow to pink, orange and red.

PROPAGATION METHODS

Easiest: *Division*. Plants grow from rhizomes. Lift the rhizomes in spring, if these plants are growing outside, and cut into pieces. If you lift the rhizomes and store them for the winter, divide them when you replant them in the spring.

Additional methods: *Seed*. Plant seeds in early spring. Nick the seedcoat, and soak it for 24 hours before planting it in a deep tray filled with a humus-rich seed starting mix. Keep the soil around 21° to 24°C using a heating mat. Seeds will germinate in 3 weeks to 2 months.

Potential Problems

To survive on their own, divisions must have a prominent "eye" where a stem, leaves and flowers will grow. Examine the rhizomes carefully before you cut them into pieces. When scarifying the seeds, make only a shallow nick in the seedcoat; don't cut into the interior of the seed.

Chaenomeles spp.

Flowering Quince, Japanese Quince
ROSACEAE

Zones: 5–9

The three species in this genus come from China and Japan. These large shrubs or small trees are grown for their early spring flowers and fall fruit. They are not self-fertile; two varieties are needed to produce fruit. The white, pink, or red blooms coat the tree, both before and after the leaves enlarge. The tart fruits are edible when cooked and used in preserves or pie. Popular quinces include cultivars such as 'Enchantress', 'Apple Blossom', 'Red Chief' and 'Pink Lady', and suppliers continually introduce new cultivars, some thornless and others with increasingly attractive flowers and fruit.

PROPAGATION METHODS

Easiest: *Seed*. Seed is available from speciality companies. Plant as soon as the seed is ripe, and place the planting trays in a cold frame over the winter months. Seeds will germinate erratically in spring. Or hold the ripe seeds in the refrigerator until spring, and give them alternating day and night temperatures after you plant them in spring.

Additional methods: *Softwood cuttings*. Take cuttings in spring, before they begin to harden up, and start them in a fresh, soil-less medium.

Hardwood cuttings. In late autumn to early winter, take hardwood cuttings. Bundle and bury them in a container under the soil under a cold frame or protected garden spot for the winter.

Potential Problems

Seeds will not germinate uniformly, so it's important to keep the seedling tray for at least the season, preferably into the following season. Stray plants may surprise you as the months go by. Fungal diseases will attack if the cuttings are too moist. Protect the buried hardwood cuttings from excess moisture over the winter months by keeping them in a cold frame or covered with a protective barrier.

Canna spp.

Clematis spp.

Chelone spp.
Turtleheads
SCROPHULARIACEAE

Zones: 3–8

The six species in this genus come from North America. They are well loved because of their white, pink or purple flowers that rise above the leaves. The tall plants make a striking accent in beds and borders simply because they stand so tall and straight. In late summer, when they begin to bloom, they make an even stronger impact. Flowers are long lasting and stand up well to early autumn wind and rain.

PROPAGATION METHODS

Easiest: *Division*. Divide plants in early summer, before they bloom. Replant immediately. If divisions look as if they are struggling to become established, pinch off the flower stalks as soon as they form. By the following year, they will be strong enough to put on a good show again.

Additional methods: *Seed*. Plant seed in starting trays in early spring, and place the trays where soil temperatures will range around 13° to 18°C. They will germinate in 2 weeks to 1 month.

Softwood cuttings. Take softwood cuttings from shoots that grow at the base of the plant when they are about 15cm long. Root them in a soil-less medium. They root quickly.

Potential Problems

This plant is known for being vigorous and sturdy. However, if you take divisions that are too small or allow a struggling plant to bloom the same year you divided it, it will suffer. Give it the same care that you give to more delicate plants, and it will reward you with strong, healthy growth.

Chrysanthemum spp.
Chrysanthemums
ASTERACEAE

Zones: 4–9

The 20 species in this genus come from the Mediterranean region, Russia, China and Japan. This group is incredibly diverse, and botanists do not always agree about the genus to which various species belong. But most home gardeners are far less interested in the botanical classification than they are in the various forms of the blooms. Most garden chrysanthemums are bushy, upright plants with clusters of blooms that are characterized as semidouble, spoon, anemone, quill, spider, pompon and decorative. Other flower forms include irregular incurve, reflex, regular incurve, intermediate incurve and brush. Look for appropriate cultivars at your garden centre and in regional catalogues.

PROPAGATION METHODS

Easiest: *Division*. Divide plants in early spring, just before they begin growing vigorously. Replant immediately.

Additional methods: *Seed*. Annuals and some hardy perennial chrysanthemums are easily started from seed. Start them in early spring – at least 8 but preferably 12 weeks before the frost-free date in a medium that contains compost. Annuals germinate best with a soil temperature of 16° to 18°C, day and night, but hardy perennials respond better when their nighttime soil temperatures are decreased to about 10°C. Annuals will germinate in about 1 week to 10 days, and hardy perennials will germinate within 3 weeks.

Softwood cuttings. Take cuttings from the base of the plant as soon as shoots are about 15cm long. Root them in a moist, soil-less medium. They form roots rapidly.

Potential Problems

Divisions will die if they are too small or the soil is allowed to dry before the plants are established enough to forage for water at deeper levels. Don't make your divisions too small, and remember to keep them well watered for their first season. Seeds are prone to damping-off diseases; protect them by layering fresh vermiculite over the top of the potting soil and planting into it. Keep air circulation high in the area where seedlings are growing. Fungal diseases also attack rooting cuttings. Be sure to use fresh, clean materials for your rooting medium.

Clematis spp.

Clematis
RANUNCULACEAE

Zones: 3–9

The more than 200 species in this genus come from diverse locations, including Europe, China, Australia and North and Central America. These plants are grown for both their habit – most are tall, climbing vines with deep green, glossy leaves – and their flowers. They are generally divided into three groups, according to the season in which they bloom, which influences the season in which they need pruning. Flower forms include large singles, large doubles, saucer shaped, star shaped, open bell shaped, bell shaped, tulip shaped and tubular. Favourite clematis include *C.* 'Jackmanii', with large, single, purple blooms; *C. florida* 'Sieboldii', with large white flowers with a green, white and maroon centre; *C. tangutica*, with bell-shaped yellow blooms and striking, glossy seedheads; and *C.* 'Nelly Moser', with pink-striped, large, white, single blooms.

PROPAGATION METHODS

Easiest: *Layering.* Plants layer themselves naturally, so all you have to do to encourage this is select the vine you want to propagate, wound the bark, pin it to the soil surface and wait for it to root.

Additional methods: *Division.* Divide late-flowering plants in early spring and spring-blooming plants in autumn.

Softwood cuttings. Take softwood cuttings from the new growth of the previous season's vines or, preferably, as basal cuttings of new vines emerging from the crown. These cuttings should be 10 to 15cm long.

Seed. Seeds are available for species clematis. Plant them in autumn, covering them with a scant layer of vermiculite. Place a pane of glass over the seed tray, and put it in a cold frame or protected spot on the north side of the house for the winter. In spring, bring it into a greenhouse or propagating area inside, and place it on a heating mat set at 21° to 24°C.

Potential Problems

If you layer a clematis vine, you can expect that it will form roots and be ready to transplant a year from the time you begin. The only problems occur when the soil dries. Remember that you are propagating the vine, and keep it moist during dry periods. Divisions are also reliable as long as they are large enough and you keep them moist while they are becoming established. Cuttings can be attacked by fungi. To speed up the rooting process and therefore decrease the amount of time the cutting is vulnerable, place the rooting medium on a heating mat set for 24°C. Seeds germinate erratically. Some seeds will germinate the first spring, but others will germinate sporadically through the summer and into the following spring.

Clethra spp.

Summersweets, Sweet Pepperbushes, White Alders
CLETHRACEAE

Zones: 5–9

The more than 60 species in this genus come from Asia and North America. They are grown for their beautiful and fragrant flowers as well as their lovely arching habit. Evergreen species, such as the lily-of-the-valley tree (*C. arborea*), make good accent or specimen plants in the garden, as do some of the deciduous species, such as *C. barbinervis*. Its lovely peeling bark makes a dramatic show once the leaves have fallen.

PROPAGATION METHODS

Easiest: *Suckers.* Plants naturally sucker, so this is a simple way to propagate them. Dig and sever from the parent plant in early spring.

Additional methods: *Seed.* Seeds are available from speciality seed companies. Plant seeds in either spring or autumn, and keep the seed trays at temperatures of 7° to 10°C. If you plant seeds in autumn, set them in a cold frame for the winter months and bring them into a protected inside spot in spring.

Semiripe cuttings. Take cuttings once they have begun to harden up at the base, generally

in early to midsummer.

Potential Problems

Suckers will not survive without an adequate root system. Brush soil away from the root area to make sure that the roots are large enough to sustain the top growth before you sever the sucker from the parent plant. Seeds germinate erratically; keep the seed tray moist for at least a year after planting. Cuttings are susceptible to fungal infections. Use only fresh, clean media for rooting, and keep air circulation high.

Convallaria spp.

Lilies-of-the-Valley
LILIACEAE

Zones: 2–7

The three species in this genus come from northern temperate areas. They are grown for their stalks of nodding, bell-shaped, white flowers and clear green, oval leaves that seem to grow directly from the ground. The flowers have a sweet, pervasive fragrance that travels well on springtime breezes, and the plant thrives in shady spots or in a location where there's filtered sun. *C. majalis* 'Aureovariegata' has pale yellow to cream-colored stripes on its leaves, 'Variegata' has white splotches on its leaves, and *C. majolis* var. *rosea* has pink flowers.

PROPAGATION METHODS

Easiest: *Division.* Plants grow from rhizomes, generally called pips, that are easy to separate once they are the correct size to grow on their own. Dig up the rhizomes in autumn and gently separate them before replanting.

Additional methods: *Seed.* Save seeds or buy them from a supplier. If you save them, you'll need to remove them from the fruit that surrounds them before planting. Suppliers normally do this step for you. The seed and surrounding flesh will give you an upset stomach if eaten, so wash your hands well after handling any. Plant the seeds in a tray filled with a humus-rich soil mix, and set the tray in a cold frame or protected spot in the garden for the winter. Seeds will germinate in spring.

Potential Problems

Divisions must have both eyes where leaves will emerge and roots develop in order to survive. Examine the rhizomes carefully before separating them. Seeds must be kept moist and cool over the winter months. Don't allow the cold frame to overheat or let the seed tray dry out.

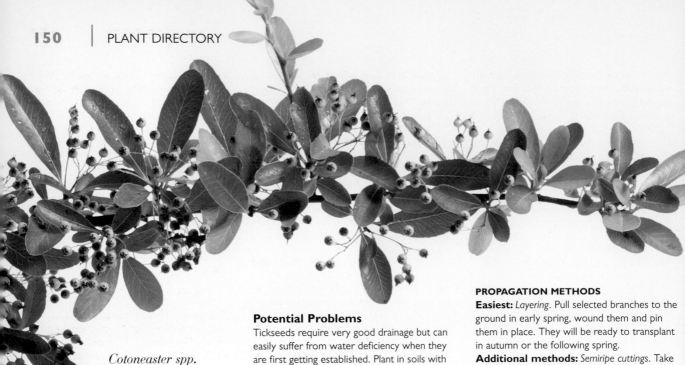

Cotoneaster spp.

Coreopsis spp.
Tickseeds
ASTERACEAE

Zones: 4–9

The nearly 100 species in this genus come from Mexico and North and Central America. They are grown for their cheery flowers, which are bright yellow and single in most species. There are both annual and perennial species, but all have long blooming periods and, if started early enough, flower their first year. *C. grandiflora* 'Early Sunrise' has semi-double blooms, 'Moonbeam' has threadlike leaves so slender that from a distance flowers look as if they are floating on air, and *C. rosea* has pink flowers.

PROPAGATION METHODS
Easiest: *Division*. Divide perennial plants in early spring.
Additional methods: *Seed*. Seeds for both annual and perennial species are available. Start them in very early spring. They germinate best at soil temperatures of 13° to 18°C. They will germinate in about 3 to 4 weeks. You can transplant annuals when they are 6 weeks old, but don't transplant perennials until they are 8 weeks old.

Basal cuttings. Snip off new stems growing from the crown when they are 10 to 15cm long. Root them the same way you do softwood cuttings.

Potential Problems
Tickseeds require very good drainage but can easily suffer from water deficiency when they are first getting established. Plant in soils with fast drainage, but keep it moist while the divisions are newly planted. Seeds will not germinate well in soils that are too warm; make certain to start seeds in an area where you can keep their trays cool. Basal cuttings root quickly enough so that problems with fungal infections are rare. However, you may have to protect against water loss from the leaves. Remember to strip off the bottom leaves when you stick the cuttings in the rooting medium.

Cotoneaster spp.
Cotoneasters
ROSACEAE

Zones: 5–8

The 200 or so species in this genus come from northern temperate areas in Europe, Asia and Africa. They are grown for the profusion of white or pink blooms covering the branches in spring and the red or yellow berries in autumn. Some species form mats growing only a few inches above the ground, but most are shrubs about 1.5 to 3 m tall and wide. Most are evergreen or semi-evergreen, but some are deciduous. Hundreds of cultivars are available, and new ones are introduced every year. Favourites include *C. franchetii*, with grey-green leaves, white flowers flushed with red, and bright orange autumn berries; *C. horizontalis*, a deciduous species with forked branches and bright red leaves in autumn; and *C. salicifolius*, a 4.5-m-tall and -wide floriferous shrub studded with shiny red autumn berries.

PROPAGATION METHODS
Easiest: *Layering*. Pull selected branches to the ground in early spring, wound them and pin them in place. They will be ready to transplant in autumn or the following spring.
Additional methods: *Semiripe cuttings*. Take cuttings when they are ready, from mid-spring to late summer.

Seed. A surprising number of cotoneasters can be grown from seed. If you collect your own, remove them from the berries, and either plant them immediately in trays destined for the cold frame or a protected spot in the garden for the winter or package them in damp peat moss and store them in a plastic bag in the refrigerator. Let them experience natural, alternating day and night temperatures in the spring. They will germinate erratically over the period of a month or two.

Potential Problems
Layered plants should pose no problems unless they are severed from the parent plant before they have formed a root system that can adequately sustain them. Check to be sure that the root system is well developed before transplanting a layered plant. Seeds germinate erratically. Keep the tray well watered and protected from overheating and excess rain for at least a full season.

Cypripedium spp.
Lady's Slipper Orchids
ORCHIDACEAE

Zones: 2–7

The 45 species in this genus come from temperate regions of the northern hemisphere, southern Asia and Mexico. They are grown for their spring flowers – dainty orchids growing only a few centimetres above the ground – and are also found growing wild in many wooded areas. Flower colours include yellow, pink, white, red and purple, and the lower lip of the flower is often shaped like a postman's bag or the "slipper" of its common name.

PROPAGATION METHODS

Easiest: *Division.* Divide their rhizomes in early or mid-spring. Replant immediately, taking as much of the soil around the roots as possible and disturbing the plants minimally.

Additional methods: *None.* Division is the only reliable way to propagate these plants on a home level.

Potential Problems

The roots require mycorrhizae, the beneficial fungi that grow in association with them. These fungi are in the soil around the roots, so you're bound to transplant them along with the orchid if you disturb the roots as little as possible. In addition to being certain to take some soil along with the division, you must be careful to keep the plants moist while they are becoming established.

Dahlia spp.

Dahlias
ASTERACEAE

Zones: 8–11

The 30 species in this genus come from Central America. They are grown for their flowers, which are so diverse that the genus is separated into categories according to flower size and then into 11 classes according to flower form. They grow from tubers that are easy to lift and store for the winter months. All species, even "giant" forms, bloom by midsummer when planted after the threat of spring frosts is past. It's difficult to pick favourites among the literally thousands of available cultivars, but old standbys such as 'Zorro', 'Jessica', 'Charlie Two' and 'Claire de Lune' continue to be garden staples.

PROPAGATION METHODS

Easiest: *Division.* Tubers reproduce naturally, and you can usually find small tubers when you dig plants for the winter. Store these as you do larger ones, and plant them out in spring. They

may take a couple of years to bloom, but eventually they will reward your care. You can also cut large tubers into pieces, each with one or more "eyes".

Additional methods: *Basal cuttings.* To take these cuttings, start the tubers in a greenhouse or well-lit area of the home in very early spring. Once shoots growing from the tubers are about 10cm tall, cut them off and stick them into a rooting medium. Place their rooting containers in a propagation chamber with both mist and bottom heat that maintains temperatures about 24°C. They will root within 2 weeks at the latest.

Seed. Seed for small bedding dahlias is widely available. Start them about 6 to 8 weeks before the frost-free date in your area, and set their starting containers over bottom heat so the medium retains temperatures between 18° and 21°C while they are germinating.

Potential Problems

Tubers cannot survive if they do not produce stems and leaves; make certain that all pieces of any tubers you cut have at least one eye. If you cut or break tubers before you store them, they are susceptible to rotting, so store them as they come out of the ground. Cuttings require bottom heat and mist to root quickly. If you can't provide these conditions, you may

still be able to root cuttings by enclosing them in a plastic bag and setting them over bottom heat. However, your chances of success are diminished. Seeds are close to foolproof, but remember that the seedlings are extremely tender. Hold them inside until there is absolutely no chance of spring frosts, and gradually harden them to outside conditions.

Delphinium spp.

Delphiniums
RANUNCULACEAE

Zones: 4–8

The 250 species in this genus come from every continent except Australia and the arctic regions. They are grown for their tall spikes of elegant flowers that bloom in white, pink, purple, blue, yellow and red. Flower spikes can be tightly packed and cone shaped, as they are in the garden favorites of 'Black Knight' and 'Astolat', or more loosely arranged on the spike, as they are in *D. cardinale* and 'Blue Butterfly'. Because delphinium can be short-lived, it's wise to regularly propagate favourites.

Dahlia spp.

PROPAGATION METHODS

Easiest: *Seed.* Stratify the seeds for at least 2 weeks in the refrigerator before planting in a humus-rich soil mix. Seeds require dark for germination; after lightly covering them with soil, put a thick layer of newspapers or cardboard on top of the tray. Place the tray where the soil will remain at temperatures of 10° to 13°C. Seeds germinate within 2 weeks.

Additional methods: *Division.* Divide plants in early autumn, and replant immediately.

Basal cuttings. It's possible to root young shoots emerging from the crown. Take them when they are 10 to 15cm long, and root them in a propagating chamber that maintains temperatures of 16° to 18°C.

Potential Problems

Seeds go dormant if subjected to high temperatures, so it's important to hold the seed trays where they simply can't overheat. This is made easier by the seeds' requirement for darkness – they do well in a cool basement rather than a hot utility room. Cuttings are prone to fungal infections. Use fresh, clean soil-less media to minimize problems and keep air circulation high.

Divisions will fail if they do not become established by the time the ground freezes. Divide and replant at least 40 days before your first expected autumn frost.

Dianthus spp.

Pinks, Carnations
CARYOPHYLLACEAE

Zones: 3–9

The 300 species in this genus come from Europe and Asia. They are grown for both their foliage, which in species such as maiden pinks (*D. deltoides*) forms a dense green mat of slender leaves, and their flowers, which range from the tall, perpetually flowering carnations found in commercial bouquets to the lovely sweet William (*D. barbatus*) blooms that get the summer border off to a colourful start. This diverse genus includes annuals, biennials and perennials, and flowers can be fragrant or not. Flower colours are white, pink, red and yellow, but the striking features are the stripes, picotees and target patterns on the petals.

PROPAGATION METHODS

Easiest: *Seed.* Seed for all types of pinks and carnations is widely available. Plant inside in early spring, and keep seed trays at temperatures of 16° to 21°C. Seeds will germinate in 2 to 3 weeks.

Additional methods: *Layering.* Tall perennial plants can be layered. As soon as you can bend a stem without breaking it, make a small wound where it will come into contact with the soil, and pin it to the soil surface. If practical, layer it in a pot sitting next to the plant because you won't have to bend the stem so close to the ground. It will root within a month. Sever the stem from the parent, and grow it on in the pot until early autumn, when you can transplant it to the garden.

Potential Problems

Seeds are close to foolproof, but seedlings of some species are quite tiny. Consequently, they are easy to kill with poor watering practices; mist them with a hand-mister held far enough away so that they never feel a blast of air or water. Layered plants will die over the winter if they are not well enough established to withstand the freezing and thawing action that is normal in most locations. If you don't think the plant has had time to get established in time for winter, sink the whole pot in the soil. In spring, you can dig it up and transplant the young plant to the garden.

Dicentra spp.

Bleeding Hearts, Dutchman's Breeches
FUMARIACEAE

Zones: 3–9

The more than 20 species in this genus come from Asia and North America. They are grown as much for their delicate foliage and bushy plant habit as for their long panicles of heart-shaped flowers. Flower colours include white, yellow and pink, and blooms are often bicoloured. Plants range in size from 36cm to 1.5 m tall, but the most commonly seen species, such as fringed bleeding heart (*D. eximia*), are about 60cm tall and wide.

PROPAGATION METHODS

Easiest: *Seed.* Seed for many species of bleeding heart is widely available. Stratify it for 2 to 3 months before planting it on the surface of the seed tray. Fill the tray with a compost-based mix, and add a layer of vermiculite to seed into. The seeds need light to germinate and may take as long as 3 months to do so. Consequently, it's best to cover the tray with plastic film and set it where the soil temperature will remain around 13° to 16°C.

Additional methods: *Division.* Divide plants in early spring, before you can see much new growth, and replant immediately.

Potential Problems

Seeds will not germinate if they do not experience a prolonged chilling period or if they overheat while they are in the seed tray. However, if you satisfy both these conditions, germination is excellent. The root system of divisions must be large enough to supply the top growth while the plant is getting established. Don't divide them into pieces that are too small, and keep them well watered after you have replanted them.

Echinacea spp.

Echinacea spp.
Coneflowers
ASTERACEAE

Zones: 3–9

The nine species in this genus come from North America. They are grown primarily for their flowers, which bloom in shades of white, pink and purple. The roots of purple coneflower (*E. purpurea*) are said to have medicinal qualities, so some people grow the plant to make tinctures. Favourite garden cultivars include 'White Swan', with white flowers and deep orange centres, and 'Magnus', with huge deep purple flowers with dark orange centres.

PROPAGATION METHODS
Easiest: *Seed.* Start seeds early inside, and keep the seed tray at about 13° to 16°C. They will germinate in 2 to 3 weeks.

Additional methods: *Division.* These plants do not tolerate a lot of root disturbance, so it's best to divide them as soon as the ground has thawed enough to dig. Rather than lifting the entire root ball, expose it on one side. Cut through the root ball and, working carefully, lift only that portion of the plant. Immediately fill in the hole with soil and replant the division. Mulch it well to retain moisture over the first season.

Root cuttings. You can take root cuttings in autumn. Do this without digging the plant. Again, simply expose a part of the root ball. Lay the root cuttings horizontally in the rooting medium rather than standing them upright, and cover them with about 0.5cm of medium. Place the rooting medium in a cool location in the house. They will root and send up shoots by spring. Alternatively, you can bury the cuttings in the soil under a cold frame to overwinter outside.

Potential Problems
Seeds will not germinate well if overheated; maintain cool temperatures for the best results. As explained above, divisions can be tricky. However, if you take proper precautions, you should have no trouble. Root cuttings are as susceptible to fungal infections as any other cuttings. Make certain that the medium is sterile when you place the cuttings in it, and cover it tightly with plastic wrap as soon as they are buried in it.

Eranthis spp.
Winter Aconites
RANUNCULACEAE

Zones: 4–9

The seven species in this genus come from Europe, Asia and the Far East. They are grown for their lovely little flowers that are often the first blooms you see in spring. The flowers of most species and cultivars are a clear, bright yellow, but *E. pinnatifida*, an alpine plant, has white flowers. The plants naturally form large masses if the leaves are left uncut until they wither. They prefer growing in a semishaded, damp area with alkaline to only slightly acid soil.

PROPAGATION METHODS
Easiest: *Division.* Tubers naturally increase, and plantings will enlarge every year. To separate the plants and give them more room or move some to another spot in the garden, separate and transplant the tubers just after the plants bloom – don't wait until autumn, as you do with many springtime bulbs.

Additional methods: *Seed.* Speciality seed companies sell seed for winter aconite, and you can also collect your own. As soon as the seed is ripe, plant it on the soil surface in a small tray, enclose the tray in plastic wrap and then a closed plastic bag, and put it in the refrigerator. Leave it there for at least 3 weeks to 1 month. Then take it out and put the tray in a cold frame. Some seeds may germinate right away; some may wait until the following spring. Prick out seedlings after they have their first true leaves, but do not disturb the rest of the tray. Grow seedlings in containers sunk into the soil under the cold frame if plants are not large enough to transplant by autumn.

Potential Problems
Winter aconite does not fare well if its tubers dry out at any time. When buying tubers, buy only those that are packed in moist peat, otherwise they won't sprout. Remember this when making your divisions, too. Quickly move them – roots, tuber, shoots and all – to moist soil when you divide. Seed germination is really erratic; don't count a tray "dead" for at least 18 months after planting it.

Eryngium spp.
Sea Hollies
APIACEAE

Zones: 3–9

The more than 200 species in this genus come from Europe, North Africa, Turkey, Asia and Korea. They are grown for their unusual conelike, dry, blue-green, lavender-green, or grey-green flowers with prominent surrounding bracts. The blooms are long lasting in the garden and make a strong statement in any border. However, they can make an even stronger statement in

Eryngium spp.

a dried arrangement that can take center stage in your home for the whole winter. Cultivation requirements vary, depending on the species. For example, the popular 'Oxford Blue' cultivar of *E. bourgatii* requires dry soil with moderate to low fertility levels. In contrast, 'Silver Ghost', a popular cultivar of *E. giganteum*, requires moist soil with high fertility levels.

PROPAGATION METHODS

Easiest: *Seed.* Plant ripe seed in containers, and set them in a cold frame for the winter. Don't cover the seed with medium; they must see light to germinate. They must also experience a chilling period, which they will certainly get in the cold frame. Alternatively, you can handle them by putting the plastic-wrapped tray in the refrigerator for at least 3 weeks to 1 month. Then take it out and put the tray in a cold frame. Some seeds may germinate right away, and some may wait until the following spring. Prick out seedlings after they have their first true leaves, but do not disturb the rest of the tray. If seedlings are not large enough to survive on their own over the winter, sink their containers into the soil under a cold frame.

Additional methods: *Division.* Divide plants in early spring. Be very careful when you do this – they resent disturbance.

Root cuttings. Take root cuttings in autumn, pot them up in rooting medium and leave their pots in the cold frame for the winter.

Potential Problems

Seed from seed suppliers may not have been properly stratified. Avoid disappointment by taking half the seed in the packet, planting it in a tray and stratifying it as specified above (but rather than putting the tray in a cold frame after it comes out of the refrigerator, put it on a heating mat set to maintain soil temperatures at 18° to 24°C). Plant the other half of the seed packet as you ordinarily would. To divide plants well, your timing must be good. Divide them as early in the spring as you can sink a spade in the soil. Keep them well mulched throughout the season, and protect them against bright sun if they have not become well established by midsummer. Root cuttings are prey to all the pathogenic soil-dwelling fungi. Use fresh, clean media to minimize chances of infection.

Erythronium spp.
Dog's Tooth Violets
LILIACEAE

Zones: 3–9

The more than 20 species in this genus come from Europe, Asia and North America. These plants are grown for their stunning spring flowers as well as their bloom time – early spring – and tolerance of deciduous shade. The flowers do not look like violets but like miniature lilies. They hang down in groups from slender stems and range in colour from white to pink, mauve, violet and yellow. The stamens of all species are prominent, as they are in many lilies. Leaves are formed in pairs and are green mottled or striped with yellow or a colour best described as dull maroon.

PROPAGATION METHODS

Easiest: These plants reproduce quite well naturally. Because they are bulbs, they produce offsets every season. If they like the environment – fertile soil and filtered light or dappled shade – they will quickly expand their area. If you want to help them along, dig them up in autumn and replant immediately. When buying bulbs, buy only those packed in damp peat moss or a similar material – this bulb cannot tolerate being dry at any time.

Additional methods: *Seed.* Speciality companies sometimes supply seeds for dog's tooth violet plants, or, more practically, you can save your own. They require chilling, so either plant them in a tray that you keep in the refrigerator for the winter months or set the planted tray in a cold frame. In spring, place the flat where soil temperatures will range around 10° to 16°C. They may take a full year to germinate.

Potential Problems

The only possible problems you could have when lifting and dividing bulbs would be letting the bulbs dry too much before replanting them or splitting them with your digging tool. Impatience is likely to be the biggest problem when starting these from seeds. Remember to keep their medium moist and prevent it from overheating – shade it in summer, if necessary. Young plants will reward this care.

Eucalyptus spp.
Gums, Ironbarks
MYRTACEAE

Zones: 8–10

The more than 500 species in this genus come from Australia, the Philippines, Malaysia, Indonesia, Papua New Guinea and Melanesia. Although they are tropical and subtropical plants, many gardeners in northern areas grow them as potted specimen plants, either indoors year round or moved to the outside garden during the summer and then back inside again in winter. Gardeners value them for their aromatic foliage, which in most species is a soft grey-green, and the decorative peeling bark that ranges in colour from a rich cinnamon orange to white, grey and brown.

PROPAGATION METHODS

Easiest: *Seed.* Seed for popular species such as lemon-scented gum (*E. citriodora*) and silver dollar eucalyptus (*E. cinerea*) is widely available and easy to germinate. Chill it in the refrigerator for 2 weeks, and then plant it shallowly. Place the starting container on a heating mat you can set for 21° to 24°C. If you are interested in more exotic species, search the World Wide Web for companies in Australia. Buy the seed, prechill it as described above, and then soak it in smoke water.

Additional methods: Seed is the only way to propagate this plant on a home level.

Potential Problems

Germination is erratic but almost always successful. Problems come when you try to grow the seedlings. Transplant them to growing containers as soon as you see the true leaves. These plants do not like root disturbance, so you want to minimize it by moving them when they are small. Always grow them in a container with smooth sides so you can easily move them into larger pots – you don't want to have to struggle with the root ball. Second, you may need to stake them when they are young to prevent them from falling over. Tie them loosely, however; swaying in the wind will help to strengthen their trunks.

Euonymus spp.
Spindle trees, Euonymus
CELASTRACEAE

Zones: 4–9

Most of the 175 species in this genus come from Asia. They are now grown all over the world because of their ornamental foliage, which turns to vivid colours in the autumn, and their decorative fruit. Euonymus can be evergreen, semi-evergreen and deciduous and grow as shrubs, trees and climbing plants. Leaves range in summer colouration from solid green to green with yellow stripes and with variegations of white, cream, or yellow. Most species turn bright red in autumn, and fruits are also vividly coloured and open to reveal the seeds. Burning bush (E. alatus) is one of the most commonly grown species, as are cultivars of winter creeper (E. fortunei) and Japanese spindle tree (E. japonicus).

PROPAGATION METHODS
Easiest: *Layering.* This method is almost always successful. Support a pot of soil close enough to the plant so a branch will easily bend to it. Wound the bark, pin it down and let it root. If it is not ready to be transplanted before winter, sink its container into the garden soil and transplant it in the early spring.

Additional methods: *Semiripe cuttings.* Take cuttings in summer, as soon as they have begun to harden up, and root as usual.

Greenwood cuttings. These cuttings are recommended for deciduous cultivars. Root them in a fresh, clean medium, preferably in a propagation chamber.

Potential Problems
Fungal diseases can attack cuttings, but if you use fresh media and hold the cuttings in an appropriate environment, they will probably be safe from infection.

Eustoma spp.
Lisianthus, Tulip Gentians
GENTIANACEAE

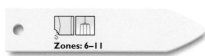

Zones: 6–11

The three species in the *Eustoma* genus come from Central and South America and the southeastern United States. These plants are grown for their stunningly beautiful flowers that last both in the garden and in the vase. You can find single and double cultivars, ranging in colour from pale yellow to pink and blue. Some have picotee markings on their petal margins, but others are a solid colour. Once relatively unusual garden plants in zones colder than hardiness Zone 8, they are becoming common as gardeners discover them. Favourite cultivars include those in the Heidi and Echo series, 'Maurine Blue', and the heat-resistant Flamenco series.

PROPAGATION METHODS
Easiest: *Seed.* Start seeds very early in spring because they are slow growing. They generally germinate in 2 to 3 weeks but take at least 8 to 10 weeks to reach transplant size. They require light and soil temperatures of 21° to 24°C to germinate.

Additional methods: *Division.* In areas where these plants are hardy, you can divide them in early spring, before they begin vigorous new growth.

Potential Problems
Seeds are vulnerable to damping-off organisms, partially because they do take so long to grow much above the surface of the medium. Minimize trouble by planting them in a thin layer of vermiculite over the starting medium and making certain that the medium is fast draining. Keep air circulation high in the area where they are growing. Divisions are sturdy; you won't have problems if you replant them in an appropriate location.

F

Ficus spp.
Figs
MORACEAE

Zones: 8–11

The approximately 800 species in this genus come from subtropical and tropical regions all over the world and range from trees to shrubs to climbing plants. Almost all are evergreen. They are grown for their impressive foliage and, in some cases, their massive trunks or their fruit. Weeping fig (F. benjamina) is a popular indoor plant in temperate areas, as is the rubber tree (F. elastica). The common fig (F. carica)—which produces the edible fig—is one of the hardiest of the genus, surviving with winter protection in Zone 6.

PROPAGATION METHODS
Easiest: *Air layering.* This is the safest way to propagate rubber trees. Begin the air-layering process in mid-spring to midsummer, when the plant is growing vigorously.

Additional methods: *Semiripe cuttings.* This method is preferable for species such as common and weeping figs. Take the cuttings in mid-spring to midsummer, as soon as the wood has begun to harden up.

Seed. Seed for some cultivars is available from speciality companies. Plant it in spring. It requires light for germination, so place it on the surface of a vermiculite layer over a compost-based soil mix, and mist it into niches in the vermiculite. Cover the flat with a pane of glass or plastic wrap to create a consistently moist environment, and place it on a heating mat set for 18° to 21°C.

Potential Problems
Air layering requires a warm environment but not a hot one. Remember to wrap aluminium foil over the air layer if the plant is in a location where sunlight will overheat the medium under the plastic covering. Semiripe cuttings are susceptible to fungal infections. To speed the rooting process, set the rooting container on a soil heating mat set to temperatures around 18° to 21°C. Seed will die if it overheats, too. Place it where it receives light from a set of fluorescent tubes or diffused sunlight, but not in direct sun.

Forsythia spp.
Forsythias
OLEACEAE

Zones: 5–9

Six of the seven species in this genus come from eastern Asia, and one species comes from Europe. They are often grown for the bright yellow, four-petalled flowers that coat the branches in early spring, before the leaves break bud, and they also make excellent hedges. They can become unruly if they are not pruned; refrain from using them in a formal planting unless you know you can keep up with pruning every year right after they bloom.

PROPAGATION METHODS
Easiest: *Layering.* Pull selected branches to the ground anytime during the growing season, wound them and pin them in place. They will be ready to transplant in autumn or the following spring.

Additional methods: *Greenwood cuttings.* Take these cuttings in spring.

Semiripe cuttings. These cuttings are generally ready in mid- to late summer.

Potential Problems

These plants root extremely quickly and rarely suffer problems. However, if their rooting medium or starting container is contaminated, they can contract a number of fungal diseases.

Franklinia alatamaha
Franklinia
THEACEAE

Zones: 6–9

The one species in this genus comes from Georgia, in the southeastern United States, and can no longer be found in the wild. It is grown for its graceful white flowers with prominent yellow centres and glossy green leaves that turn red in autumn. The flowers are fragrant and bloom in late summer and early autumn, when few shrubs and small trees bloom.

PROPAGATION METHODS

Easiest: *Softwood cuttings.* Take cuttings in late spring, and root in a soil-less medium. Place in a misting propagating unit if possible; if not, use bottom heat and hand-mist the cuttings several times a day.

Additional methods: *Layering.* This can be done only when the branches are still flexible enough to reach the ground.

Seed. Plant the seeds as soon as they are ripe – generally, in very late autumn. Place the seed tray in a protected spot outside or under a cold frame for the winter. Alternatively, you can place the tray on a soil heating mat set to maintain temperatures of 10°C.

Potential Problems

Protect cuttings from fungal infections by starting them in a fresh, clean, soil-less medium and providing bottom heat. Layering is a very safe way to propagate any plant, including franklinias. The only possible problem is severing the layered plant from the parent before it has a large enough root system to sustain life on its own. Seed is not at all certain. It's possible to have good results, but it's a less reliable method than cuttings, simply because seed can take a long time to germinate.

Gaillardia spp.
Blanket Flowers
ASTERACEAE

Zones: 3–9

The 30 species in this genus come originally from North and South America. They are grown for their colourful flowers, which range from red to orange or yellow. The central disks range in colour from orange to purple, brown and red. Some cultivars have fringed petals, and the petals of others, such as *G. x grandiflora* 'Dazzler', have red petals tipped with yellow. Favourite cultivars include 'Baby Cole', 'Red Plume' and 'Kobold'.

PROPAGATION METHODS

Easiest: Seed. All types of blanket flowers, from annuals to biennials and perennials, grow well from seed. Scatter them on the surface of a compost-based medium; they require light to germinate. Place the trays on a soil heating mat if possible because germination rates are highest when the soil maintains a steady 21° to 24°C.

Additional methods: *Division.* Perennial species respond well to division. Lift and divide them in very early spring, before they have resumed vigorous growth for the season.

Root cuttings. If you want to propagate a lot of a particular cultivar in a hurry, take root cuttings in the late winter or early spring and plant them vertically in a tray or pot. Place the pot in a cold frame or a protected spot in the garden, and shield it from excess rain. New plants will appear over the late spring and early summer months.

Potential Problems

Seeds are just about foolproof – just don't plant them so deeply that they aren't exposed to light. Divisions take easily, provided they have a large enough root system to sustain their top growth and you keep them moist while they are becoming established. Root cuttings are easier than you might imagine. Make certain that the rooting medium drains quickly and easily so that the cuttings aren't exposed to standing water in the pot.

Galanthus spp.

Galanthus spp.

Snowdrops
AMARYLLIDACEAE

Zones: 3–9

The 19 species in this genus come from Europe and western Asia. They are grown for their lovely nodding blooms that are among the very first flowers to appear in spring. The fragrant, snowy-white flowers often have green tips on the petals. Cultivars of *G. nivalis*, such as 'Sandersii', have a striking yellow calyx, while 'Lady Elphinstone' has a double centre that is yellow.

PROPAGATION METHODS

Easiest: *Division.* Lift and divide the clusters of bulbs as soon as the flowers fade but before the leaves do. These plants multiply quickly so that they are likely to become crowded if you do not do this every 3 to 4 years.

Additional methods: *Seed.* You can collect seed as soon as it is ripe, and plant it in a tray filled with a compost-rich soil mix. Bring the tray inside to protect it from exposure to hot sun. Transplant the seedlings to the garden bed in autumn.

Potential Problems

Bulbs die if you allow them to dry out after transplanting them. Move them carefully, disturbing the roots as little as possible, and keep them moist throughout their first summer. Most of the snowdrops sold are hybrids, so plants that grow from your self-saved seed are not likely to look like their parents. But this is certainly not a problem; all snowdrops are lovely and bloom early – the two qualities you probably desire.

Galium spp.

Bedstraws, Sweet Woodruffs
RUBIACEAE

Zones: 5–8

The 400 species in this genus come from temperate regions all over the world. Only a few are garden-worthy. They are grown for their whorls of clear green leaves and ability to create a lovely ground cover in sun or partial shade. They have white, pinkish white, or yellowish white flowers in spring. The flowers are lovely but appear only briefly. The striking thing about the plant is its fragrance. When the leaves are cut and dried, they smell like freshly cut hay or grass, making a good sachet for a linen closet or chest of drawers.

PROPAGATION METHODS

Easiest: *Division.* Plants grow from rhizomes that are easy to divide. You can do this in autumn or spring, whichever is more convenient.

Additional methods: *Seed.* Sweet woodruff produces viable seed. Plant it when it is ripe in a compost-rich potting soil, and hold it at temperatures of 16° to 21°C.

Potential Problems

Don't allow the divisions to dry out before they have become established or they will die. If you divide them in autumn and live in an area with lots of frost heaves, it's best to mulch the divisions as soon as the soil freezes in winter. Don't let the seed overheat or it will become dormant or, in worst cases, die.

Gardenia spp.

Gardenias
RUBIACEAE

Zones: 8–10

The approximately 200 species in this genus come from tropical areas of Africa and Asia. They are grown for their waxy white, fragrant flowers and glossy green leaves. They bloom for many months; the solitary flowers never coat the plant but appear sporadically over the blooming season. They thrive in dappled or partial shade and are ideal for areas with acid soil. Of the many cultivars, available favourites include 'First Love', 'Mystery' and 'Daisy'.

PROPAGATION METHODS

Easiest: *Softwood cuttings.* Take cuttings in spring, while the wood is still soft, and root them in a propagation chamber with mist.

Additional methods: *Grafting.* Both side-wedge and apical-wedge grafts are appropriate for gardenias.

Potential Problems

These are not easy plants to propagate, so it's common to have problems. With a misting propagation chamber, you will be able to root the cuttings. Without one, it may be a challenge to keep the area around them humid enough. Create a pebble tray filled with gravel and water, and set this inside a plastic tent where you've placed the cuttings. The pebble tray will give off the essential humidity. Do not set the plant pot on the tray, however, because the rooting mix would be likely to wick up the moisture and become too wet. When grafting, be scrupulously clean, and be sure to line up the cambium layers so they match exactly.

Gentiana spp.

Gentians
GENTIANACEAE

Zones: 3–9

The 400 or so species in this genus come from temperate areas all over the world, primarily mountainous zones. The plants are grown for their large, striking flowers, trumpet shaped in most species. They have a long bloom period, and most species have blue or purple flowers, although some, such as *G. dendrologii*, have white flowers, and others, such as *G. alba*, have yellow-green flowers. Many species bloom in autumn, adding a blue note to gardens when they are predominately yellow and orange. Favourite species and cultivars include spring gentian (*G. verna*), *G.* 'Devonhall' and yellow gentian (*G. lutea*).

PROPAGATION METHODS

Easiest: *Seed.* Seed is widely available for many species and cultivars. Chill the seed for 2 months in the refrigerator, and then plant in seed trays. Cover the trays with cardboard to obscure light; these seeds require darkness to germinate. Maintain soil temperatures of 21° to 24°C while plants are germinating.

Additional methods: *Division.* Divide in early spring, and replant or pot up immediately.

Basal cuttings. Basal shoots will root well in a moist medium. Cut them when they are 10 to 15cm tall, and stick them in a moist, soil-less medium. Tent the rooting tray with plastic to maintain high humidity levels.

Potential Problems

Depending on the species, seeds germinate at any time from 2 weeks to 6 months from planting. Check the packet or catalogue information. If you have seed that takes a long time to germinate, be certain to place it where you will remember to take good care of it through the months. These plants divide well but, as always, must be the proper size in order to survive long enough to become established. They are very attractive to slugs, so be extra vigilant while they are rooting. Basal cuttings are prone to fungal attack. You can minimize possibilities of infection by starting them in sterilized containers and fresh rooting media.

Geranium spp.

Cranesbills
GERANIACEAE

Zones: 5–9

The more than 300 species in this genus come from temperate areas all over the world. Many species thrive in full sun, but others grow best in partial shade. They are grown for both their decorative foliage and lovely flowers. Flower colours include white, blue, pink, purple and magenta, and plants range in size from 15 to 46cm tall. There are hundreds and hundreds of cultivars. Favorites include *G.* 'Johnson's Blue', 'Wargrave Pink' and *G. sanguineum* var. *striatum*.

PROPAGATION METHODS

Easiest: *Division.* Divide plants in early spring; many grow from tubers, and others have a fibrous root system. In either case, they are easy to divide.

Additional methods: *Seed.* Seed for many species and cultivars is widely available. Place the seed trays in a location where the soil will maintain a temperature of 10°C.

Softwood cuttings. Take cuttings in spring, and root them in a fast-draining medium, preferably a propagation chamber. If that is impossible, place their rooting containers on a heating mat that will keep the medium at about 21° to 24°C.

Potential Problems

As always, make certain that divisions have an adequate root system to survive on their own, and keep them uniformly moist while they are becoming established. Seeds of some species germinate sporadically; don't assume that they aren't going to sprout for at least 3 months. Cuttings are susceptible to fungal infections. Use only fresh, clean rooting media.

Gerbera spp.

Gerberas, Transvaal Daisies
ASTERACEAE

Zones: 8–11

The 40 species in this genus come from South Africa, Madagascar, Asia and Indonesia. They are grown for their showy single flowers in vivid shades of red, pink, purple, orange and yellow and their mats of deeply lobed leaves. Festival series cultivars are extremely popular, as are Tempo series and Parade series.

PROPAGATION METHODS

Easiest: *Seed.* Seed is widely available. Plant it in early spring, about 8 to 10 weeks before the frost-free date. It requires light to germinate, so it's best to plant it pointed end down and leave a little of the more rounded end exposed to the light. It requires temperatures of 21° to 24°C to germinate and usually takes 2 to 3 weeks.

Additional methods: *Division.* Where hardy, divide plants in either early spring or autumn.

Basal cuttings. Take these cuttings when they are 10 to 15cm long and root them in fresh, clean media at a soil temperature of 24°C. If you don't have a propagating chamber, remember to mist them frequently during the day.

Potential Problems

Seeds are a reliable way to propagate gerberas. However, like all seeds, they do require certain environmental conditions. In this case, don't let them get cold – keep the soil medium, as well as the air around them, warm while they are germinating. Divisions are also a foolproof way to propagate them as long as the pieces are large enough to survive long enough to establish themselves. Basal cuttings are susceptible to fungal infections. Nonetheless, you will be successful if you keep them warm enough while they are rooting and use fresh, clean media.

Geranium spp.

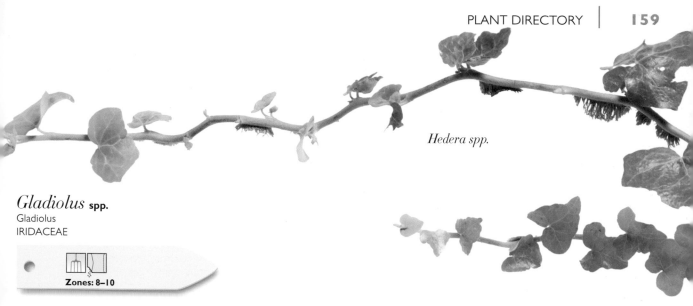

Hedera spp.

Gladiolus spp.

Gladiolus
IRIDACEAE

Zones: 8–10

The 180 species in this genus come from Africa, the Mediterranean, the Arabian Peninsula, Madagascar and western Asia. They are grown for their tall spikes of flowers that add structure to any border and are also excellent cut flowers. In most species, the flowers are tightly packed on the spike, giving a lush appearance. In species such as *G. callianthus* and *G. tristis*, however, they are widely spaced, and each droops forward from a long tube. The number of available cultivars is breathtaking and changes every year.

PROPAGATION METHODS

Easiest: *Division.* Corms reproduce themselves every year. In climates where they are hardy, you can simply leave them in place, and they will multiply naturally. In more northern areas, where you dig them up to overwinter inside, it's best to let the leaves and corms finish drying before storage. After they are dry, you can remove the old corm from the new one, which is what you want to store. You'll find some baby corms, too; store these as well and plant them in a nursery bed outside when you make your first planting of gladioli for the year.

Additional methods: *Seed.* If you have not removed the spent flower stalks, you can collect seeds from the dried flower heads in late summer and early autumn. Let them dry thoroughly, and save them until spring. Plant them in a nursery bed outside or in trays inside. Maintain a soil temperature of 21° to 24°C while they are germinating. If you live where they are not hardy, lift the plants at the end of the season, and store the baby corms that have started to form. Repeat the planting and lifting routine the following year. By the third or fourth year, your corms will be large enough to bloom. But don't expect them to look like their parent plant – they will no doubt have crossed with other gladioli.

Potential Problems

Corms are reliable. Unless you plant them in very soggy soil or before the frost-free date, you should have no problems. Gardening develops patience; gardeners have to wait years and years for some plants to become mature. Working with gladioli seed is one of these; you will only succeed if you have truly learned the lesson of patience.

Hamamelis spp.

Witch Hazels
HAMAMELIDACEAE

Zones: 5–9

The five species in this genus come from eastern Asia and North America. They are grown for their fragrant flowers that emerge at the end of winter or very early in spring, as well as for their value as an architectural feature in the garden and their vivid autumn foliage. The blooms are yellow or orange and brighten up the winter doldrums with their promise of the spring to come. The fragrance is refreshing rather than sweet.

PROPAGATION METHODS

Easiest: *Layering.* Layer branches in mid-spring, and leave them in place for the entire season or even longer – perhaps into the next year.

Additional methods: *Air layering.* Gardeners rarely air-layer plants in the garden, but it's an excellent way to propagate some plants, witch hazel among them. Do this as you would inside, taking particular care to wrap the layered area well with aluminium foil.

Softwood cuttings. Take cuttings in late spring, and root them in a propagation chamber. If that is impossible, tent the rooting pot with plastic to retain humidity.

Side-veneer grafting. Graft in early spring.

Potential Problems

Layers are close to foolproof if you leave them in place long enough so they develop a good root system. Don't be in any hurry to sever them from the parent plant. Air layers are also reliable – unless they dry out or overheat. Check the material inside the foil and plastic regularly to make sure it's sufficiently moist and not overheating. Cuttings are always susceptible to fungal infections, but clean, fresh media and containers minimize problems. Grafts fail because the wood dries out or the cambium layers aren't in contact. Make certain that the graft union is well wrapped to retain moisture and that the cambium layers are touching each other.

Hedera spp.

Ivies
ARALIACEAE

Zones: 4–9

The approximately 10 species in this genus come from temperate regions all over the world, particularly Asia, North Africa, the Canary Islands, the Azores, western Europe and Asia. They are grown for their foliage and plant habit – they are climbers and trailers and add beauty and atmosphere anywhere they grow. Leaf shapes are diverse and foliage may or may not be variegated. Mature plants have berries in autumn. Favourites include cultivars of English ivy (*H. helix*), such as the variegated 'Calico' and 'Telecurl', and cultivars of *H. nepalensis*, such as 'Suzanne' and 'Marbled Dragon'.

PROPAGATION METHODS

Easiest: *Layering.* Young plants in active vegetative growth produce adventitious rootlets along their nodes.

Additional methods: *Semiripe cuttings.* Take cuttings of young growth in summer, and root them in a misting propagating chamber or at least with bottom heat. Depending on your region and the root system of the new plant, overwinter them under a cold frame or inside the house, and transplant them to the garden in spring.

Hardwood cuttings. Take cuttings in early winter, and hold them in a cold frame over the winter months.

Potential Problems

Ivy has two distinct growing stages: the juvenile, or vegetative, state and the mature, or flowering and fruiting, state. Propagate only with the juvenile growth; mature growth is resistant to rooting, either from the nodes or meristematic tissue in the stem. Expect rooting to be slow, whether you are layering the plant or working with cuttings. Protect plants by introducing fresh air into the propagating unit and making certain that containers have been sterilized and rooting media are fresh and clean.

Helleborus spp.
Hellebores
RANUNCULACEAE

Zones: 4–9

The 15 species in this genus come from Europe and western Asia. They are grown for their stunning, saucer-shaped flowers that rise above the decorative leaves on succulent-looking stems in late winter and early spring. Species differ in their environmental conditions, so it's important to learn whether the type you are buying prefers alkaline or acid soil, sun or shade. Outstanding new cultivars are being introduced every year. Some of the favourite species and cultivars include Christmas rose (*H. niger*), with white flowers flushed pink and with a greenish centre; *H. x hybridus*;

hybrids of *H. orientalis* and other hellebores that range in colour from white to green, yellow, pink and purple; and some of the green hellebores, such as *H. cyclophyllus* and *H. argutifolius*.

PROPAGATION METHODS

Easiest: *Division.* Divide plants in spring or early summer, after they have finished flowering. *H. foetidus* and *H. argutifolius* do not divide; you'll need to start them from seed.

Additional methods: *Seed.* Plant seed as soon as it is ripe. Do not expect it to germinate for at least a year; some seeds take 2 years. Place the seed tray in a protected spot in the garden or a cold frame.

Potential Problems

It's important that the seeds don't dry out, so most people cover the tray with a pane of glass or sturdy plastic film. However, remember that the soil can overheat under this covering, so keep the tray shaded. Remember to check it for moisture levels, too.

Hemerocallis spp.
Daylilies
LILIACEAE

Zones: 4–9

The 14 or 15 species in this genus come from China, Korea and Japan. They are grown for their versatility in the garden as well as their lovely flowers and attractive foliage. There are many thousands of cultivars – as long as you aren't looking for a blue daylily, you can find one with the colour and form to suit your needs: single, double, star shaped, rounded or triangular. Many of the new hybrids are tetraploids – or plants with four, rather than two, sets of chromosomes.

PROPAGATION METHODS

Easiest: *Division.* Daylilies are notoriously easy to divide. You can divide most species anytime during the growing season, but they will fare better if you divide evergreen species in spring and herbaceous daylilies in spring or autumn, not summer.

Additional methods: *Seed.* Collect seed when the capsules open. It must go through 6 weeks' chilling time, so either hold it in the refrigerator over the winter and plant it in spring or immediately plant it in a seed tray that you'll hold in the cold frame over winter.

Potential Problems

The only possible problems you could have with divisions is replanting them into soil that doesn't drain properly or placing them in

shade. If the environment is hospitable, the divisions will establish themselves. Don't expect your seed-grown plants to resemble the parent plant. For one thing, hybrids will not come true from seed; for another, if you are growing more than one cultivar or species of daylily, it's likely that any seeds you have are homemade hybrids. But don't let this stop you from growing from seed – you could have the next popular daylily hybrid! Be patient with these plants; place them in a protected spot in the garden so they can grow undisturbed for the 3 to 4 years that it will take them to reach blooming size.

Heuchera spp.
Coral Bells
SAXIFRAGACEAE

Zones: 4–8

The 55 species in this genus originate from North America and Mexico. They are grown for both their ornamental leaves and their tiny, nodding flowers that rise on tall stems above the mound of foliage. Leaves may be green, cocoa brown, or purplish and are often mottled or streaked with silver and purple. Favourite cultivars include 'Palace Purple', 'Chocolate Ruffles' and 'Pewter Moon'.

PROPAGATION METHODS

Easiest: *Division.* Divide plants in autumn. You'll need to lift and replant them every few years anyway because their crowns tend to push out of the soil, making them more susceptible to winter damage than they otherwise might be. When the crowns appear is an ideal time to divide them.

Additional methods: *Seed.* The small seeds need light to germinate, so sprinkle them very thinly on a layer of vermiculite placed over a compost-based potting mix, and mist them into niches in the vermiculite. They germinate best at temperatures of 18° to 21°C, so place them on a heating mat if necessary. Most germinate within 2 weeks, but some can take a month or more.

Potential Problems

Coral bells can be brittle, so you must be especially gentle when dividing them. They can also take their time re-establishing themselves, so it's best not to divide them into more than two pieces. Seedlings are tiny. If you use a compost-based mix under the vermiculite and plant them thinly to begin with, you can simply leave them in place until they become large enough to move without injuring them.

Hibiscus spp.

Hibiscus
MALVACEAE

Zones: 6–11

The 200 species in this genus come from warm regions all over the world. They are grown for their large, showy flowers, ornamental foliage and versatility in the garden. Flower colours include white, pink, red, yellow, purple and blue. Leaves of most cultivars are green, but a few, such as 'Coppertone,' have bronze leaves that add to their value as specimen plants. Favourite species and cultivars include common rose mallow (*H. moscheutos*) and its Disco Belle series of cultivars, Chinese hibiscus (*H. rosa-sinensis*) and rose of Sharon (*H. syriacus*).

PROPAGATION METHODS

Easiest: *Layering.* Layer plants in spring or summer, leaving them in place long enough to develop a good root system.

Additional methods: *Seed.* Seeds must be scarified and soaked for about 24 hours in warm water. Place them on a soil heating tray because they germinate best at a soil temperature of 24° to 27°C.

Softwood cuttings. Take cuttings in late spring or early summer, as soon as they are ready, and root at a soil temperature of 24°C.

Semiripe cuttings. Take cuttings in summer, and root at a temperature of 24°C.

Potential Problems

Layered plants will not give you problems if you wait to move them until they have a large enough root system and you place them in an appropriate location. Seeds grow quickly. Make certain to prick them out of the starting containers before they get root bound. If they are destined for pots rather than the garden, continue to transplant them into bigger containers before their roots are forced to circle the bottom of the pot. Cuttings are susceptible to fungal infections, but these plants root so easily – as long as the medium is warm – that you can avoid problems if you have them in a warm, humid environment while they're rooting.

Hosta spp.

Hostas, Plantain Lilies
LILIACEAE

Zones: 3–8

The 70 species in this genus come from China, Japan, Korea and parts of eastern Russia. These plants are grown for their ornamental foliage and because they tolerate shady conditions and short periods of drought. Leaves range in colour from light yellow-greens to deep greens and, finally, blue-greens and are often variegated. Flower spikes rise above the foliage in midsummer. Flowers are bell shaped and range in colour from white to pink to lavender, blue and violet. Cultivars such as 'Patriot' have blue-green leaves margined in white, while 'Emerald Tiara' has bright green leaves.

PROPAGATION METHODS

Easiest: *Division.* Divide plants in spring or late summer.

Additional methods: *Seed.* Seed of some cultivars is available from speciality seed companies, or you may collect it from plants after the capsules split. Plant it as soon as it is ripe. Sow it into a layer of vermiculite placed over compost-based mix, and keep the medium at a temperature of 10°C. Seeds will germinate in a month or two. In cool climates you'll need to overwinter them inside the house or in a greenhouse because they won't be large enough to reliably survive a rough winter.

Potential Problems

Plants divide easily, so you shouldn't have any trouble as long as you wait until they are large enough to divide and prevent them from drying out while you are moving them into their new positions. Seeds are not always reliable; you may find that the germination rate is low or that many succumb to damping-off organisms. Keep air circulation high around the seedlings to avoid fungal infections.

Hyacinthus spp.

Hyacinths
LILIACEAE

Zones: 3–8

The three species in this genus come from Asia. They are grown for their fragrant spikes of spring flowers. Flower colours include shades of white, red, pink, yellow, purple and lavender. Leaves die back a few weeks after the flowers fade. Almost all the cultivars sold today are hybrids of the species *H. orientalis*. Favourite cultivars include 'Delft Blue', 'Pink Pearl', 'Ostara', 'Innocence' and 'Jan Bos'.

PROPAGATION METHODS

Easiest: *Division.* Plants produce bulblets that can be divided from the parent bulb in fall. Plant them in a nursery bed where they can grow for the 3 to 4 years it will take them to reach blooming size.

Additional methods: *Score.* To propagate a large number of plants from one bulb, score it (see page 70).

Potential Problems

If hyacinths grow well in the environment you've provided, you'll be able to propagate them easily from bulblets. If scoring a bulb, be careful to prevent it from developing fungal diseases. Don't let its medium dry out, but don't let it be soggy either.

Hyacinthus spp.

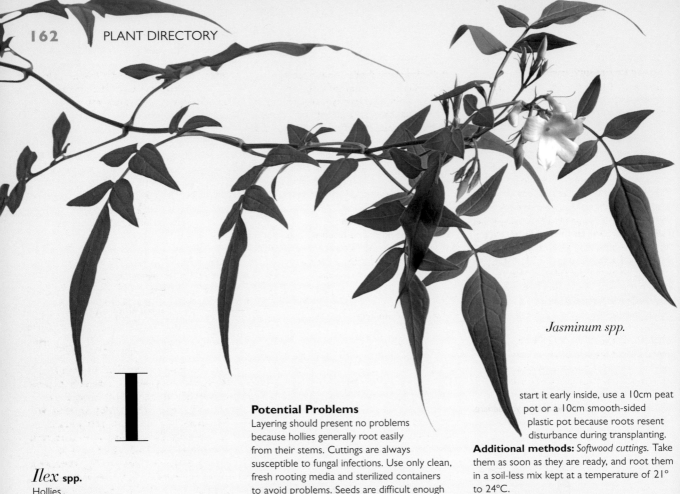

Jasminum spp.

I

Ilex spp.
Hollies
AQUIFOLIACEAE

Zones: 4–9

The more than 400 species in this genus come from temperate and tropical regions all over the world. They are grown for their decorative foliage and bright yellow or red berries. Some cultivars, such as 'Argentea Marginata' and 'Golden Milkboy', have variegated leaves, but most cultivars have deep green leaves, with or without sharp prickles on the margins. Plants are either male or female, so to get berries – and seeds – you need to grow both genders.

PROPAGATION METHODS

Easiest: *Layering*. In spring, layer plants when they are still young enough so you can bend a branch to the soil surface without injuring it.

Additional methods: *Softwood cuttings*. Take cuttings in late spring or early summer.

Semiripe cuttings. Take cuttings in mid- to late summer.

Seed. Plant as soon as it is ripe. Place the seed tray in a cold frame. Seeds can take as long as 2 years to germinate, so you'll have to keep them from overheating in summer and remember to water them.

Potential Problems
Layering should present no problems because hollies generally root easily from their stems. Cuttings are always susceptible to fungal infections. Use only clean, fresh rooting media and sterilized containers to avoid problems. Seeds are difficult enough so that the only possible reason to try to propagate a holly this way is to work on developing a hybrid between two of your favourite plants.

Ipomoea spp.
Morning Glories
CONVOLVULACEAE

Zones: 7–11

The approximately 400 to 500 species in this genus come from warm temperate, sub-tropical and tropical regions in North, South and Central America. They are grown for their flowers, which grow on vines in most species and range in colour from white, pink, red, yellow and blue to purple. Common Morning Glory (*I. tricolor*) and Cardinal Climber (*I. x multifida*) are annual; others, such as Moonflower (*I. alba*), are grown as annuals in areas where they aren't hardy.

PROPAGATION METHODS

Easiest: *Seed*. All species grow well from seed, but you'll need to scarify it and then soak it for 8 hours before planting. If you want to start it early inside, use a 10cm peat pot or a 10cm smooth-sided plastic pot because roots resent disturbance during transplanting.

Additional methods: *Softwood cuttings*. Take them as soon as they are ready, and root them in a soil-less mix kept at a temperature of 21° to 24°C.

Potential Problems
These plants are very easy to propagate, either by seed or softwood cutting. The only possible problems are fungal infections. Make certain that the seeding medium maintains a temperature of 24°C and that you keep air circulation high around the cuttings.

Iris spp.
Irises
IRIDACEAE

Zones: 3–10

The 300 or more species in this genus come from the northern hemisphere. They are enormously varied – some grow from rhizomes, some from bulbs and some have fleshy roots. Their flower forms are even more varied. You probably think of the tall bearded iris when you think of these plants, but the genus also includes beardless forms such as Siberian iris and Louisiana iris; crested, or Evansia iris; and bulbous iris such as the reticulata iris. Fortunately, propagation is similar, regardless of type.

PROPAGATION METHODS

Easiest: *Division.* Both rhizomes and bulbs are easy to divide. In the case of iris growing from rhizomes, lift the rhizome in late summer to early autumn, but wait until autumn for those growing from bulbs. Treat fleshy roots as you do bulbs – divide them in autumn.

Additional methods: *Seed.* If you don't deadhead the flowers, they will make seeds. Plant the seeds as soon as they are ripe, and place the tray in a cold frame. They can take more than a year to germinate, so you'll have to remember to water them and keep them from overheating during summer.

Potential Problems

Bearded iris can be divided as early as midsummer, after they finish blooming. But if you divide them this early, cut the fans of leaves back so the plant will not lose too much water to transpiration while it is becoming re-established. If you don't, the plant could die. Seeds are not practical unless you are trying to create a particular hybrid, in which case you must be sufficiently attentive to maintain the seed tray for the long germination period. Otherwise, simply buy some iris that you admire.

J

Jasminum spp.

Jasmines
OLEACEAE

Zones: 8–11

The 200 or more species in this genus come from Europe, Asia and Africa. They are grown primarily for their white, pinkish white or yellow flowers, which have an indescribably sweet fragrance, but are also prized as a climbing vine that is immensely versatile in the garden. Winter jasmine (*J. nudiflorum*) is hardy to Zone 6, unlike other species in this genus. Common jasmine (*J. officinale*) is popular in gardens because it produces highly fragrant flowers from summer to autumn.

PROPAGATION METHODS

Easiest: *Layering.* Layer plants in autumn. They will be ready to transplant the following autumn.

Additional methods: *Semiripe* cuttings. Take cuttings when they are ready – usually, in early spring, but this may vary according to climate. Root them in a misting propagating chamber if possible, and set the thermostat so the medium maintains a temperature of 24°C. If you don't have a propagating chamber, place the seed tray on a heating mat, put a pebble tray beside it, and tent both the flat and the humidifying pebble trays to retain high humidity levels.

Potential Problems

Layering should give no trouble; plants root easily. Fungal infections are always possible with cuttings. Remember to introduce fresh air into the tent a few times a day.

Juniperus spp.

Junipers
CUPRESSACEAE

Zones: 3–9

The approximately 55 species in this genus come from the northern hemisphere. They are grown for their lovely plant habits, which range from prostrate plants such as creeping juniper (*J. horizontalis*) and Bonin Island juniper (*J. procumbens*) to tall, cone-shaped junipers such as cultivars of *J. chinensis* and *J. communis*. Most have green leaves, although 'Depressa Aurea' has gold leaves in the spring that turn bronze and then green over the winter.

PROPAGATION METHODS

Easiest: *Layering.* Evergreen plants are as easy to layer as deciduous ones, but you need to remove the needles in the part of the branch that you'll pin to the soil. Wound the bark where the branch will touch the soil as usual.

Additional methods: *Side-veneer grafting.* This technique is easy to do and quite effective when used with conifers. Graft in early spring, just as vigorous growth is beginning.

Semiripe cuttings. Take cuttings in late spring to midsummer, whenever they are ripe, and root them in a propagating chamber with both mist and bottom heat at 16° to 18°C.

Hardwood cuttings. Take hardwood cuttings in late autumn or winter, and root them in a humid cold frame.

Potential Problems

Don't sever a layered plant before it has a large enough root system to live on its own. Whenever you graft, wrap the graft well enough that the wood doesn't dry. Make sure that the cambium layers are touching each other. Cuttings may get fungal infection. Use fresh, clean rooting media and sterilized containers to minimize problems.

K

Kalanchoe spp.

Kalanchoes, Mother of Thousands
CRASSULACEAE

Zones: 10–11

The 125 species in this genus come from Madagascar, Asia, Australia, Africa and tropical portions of North and South America. The plants are grown for their ability to withstand dry periods as well as for their flowers and ornamental, succulent leaves. Plant forms range from relatively small annuals to perennial shrubs, climbers and trees. *K. daigremontiana* is known as Mother of Thousands and is a common indoor plant in areas where temperatures fall below 10°C in winter. *K. blossfeldiana*, generally known simply as kalanchoe, is a winter-flowering greenhouse or conservatory plant that blooms in shades of white, yellow, pink and rose.

PROPAGATION METHODS

Easiest: *Plantlets.* Plants such as Mother of Thousands produce small plantlets, complete with tiny leaves and threadlike roots, on their leaf margins. Set pots of soil around the mother plant so that as plantlets fall (they will when large enough to root) you can rescue them and plant them upright in the media.

Additional methods: *Division.* Many kalanchoes produce offsets that can be divided from the parent plant after they have developed a distinct root system.

Seed. Seed for many species and cultivars is available from speciality seed companies. Plant it in early spring, sprinkling it thinly on vermiculite that covers a fast-draining potting mix. Mist seeds into niches; they need light to germinate. Hold the flat at 18° to 21°C for the week or two it takes to germinate.

Softwood cuttings. Take 10- to 15cm-long cuttings from tip ends and root them in a fast-draining soil-less medium with good drainage.

Keep the medium at about 24°C.
Potential Problems
The sheer number of plantlets that one plant produces tells you how many don't survive. However, if you capture the plantlets as they fall and place them upright in a fast-draining medium, you could have thousands more than you could possibly use or give away. Just make

certain that the pots don't dry out while the plants are establishing themselves, the soil mix drains freely, and you keep air circulation high to minimize the risk of damping off and other fungal diseases. Seeds germinate freely, but small plants succumb to fungal infections easily. Avoid problems by keeping air circulation high where they are growing after they have germinated and by using a very fast-draining mix. Cuttings root readily. In moist and warm medium, you won't have any trouble.

Kniphofia spp.
Red Hot Pokers, Torch Flowers
LILIACEAE

Zones: 5–9

The 68 species in this genus come from Africa. They are grown for their striking spikes of flowers that rise above the foliage. Flowers are greenish, pale yellow, bright yellow, orange and red. Some, such as *K. caulescens* and *K.* 'Royal Standard', appear to be bicoloured because florets are orange and red when young but yellow as they age, giving the appearance of a flower spike with orange or red blooms at the top and yellow ones at the bottom. There are hundreds of cultivars available, for use in a greenhouse or the summer garden.

PROPAGATION METHODS
Easiest: *Seed.* Seed for many species and cultivars is widely available. Plant the seed inside in early spring. Place the starting seed tray on a soil heating mat that keeps temperatures of 24°C, but remove it from the mat at night and place where temperatures will range from around 13° to 16°C.
Additional methods: *Division.* Divide plants in early spring, before vigorous growth has started. Replant them in moist, somewhat sandy soil and mulch them over one winter.
Potential Problems
Seeds can become infected with fungal diseases but are not likely to do so if the planting medium drains well and temperatures are appropriate. Divisions could die if placed in soil that stays soggy long after rain or irrigation; this plant needs good drainage. It also needs good nutrition and high humus levels, so spread compost around it every year.

Laburnum spp.
Golden Chain Trees
FABACEAE

Zones: 5–8

The two species in this genus come from central Europe and Asia. They are grown for their long weeping racemes of flowers that bloom in late spring to early summer. The flowers are yellow and look like those of sweet peas. The delicate leaves are decorative on their own. The cultivar 'Pendulum' has weeping branches that accentuate the pendant habit of the flower racemes, while those of 'Pyramidale' are upright.

PROPAGATION METHODS
Easiest: *Seed.* Seed for species, but not cultivars, of golden chain tree is available from speciality seed companies. Plant it in autumn, and set the planting trays in a cold frame to overwinter outside. Seeds will germinate in late spring to early summer.
Additional methods: *Apical-wedge grafting.* Graft in late winter or very early spring, just before the sap starts to run, and tape the union well.
Hardwood cuttings. Take cuttings in late autumn or early winter, and let them overwinter in the cold frame.
Potential Problems
Seeds overwintered in a cold frame are always in danger of being forgotten and left to dry out or overheat under a closed cold frame. If you remember to take care of the seed tray, you should have no trouble germinating these plants. Cuttings are always susceptible to fungal attack. Examine them carefully when you remove them from the cold frame in spring, and throw out any that seem even slightly diseased. Grafts fail because the cambium layers are not in contact with each other, the graft isn't well sealed and dries out, or a fungal infection strikes them. Take care to position the stock, and scion well and tape the union securely. The tape prevents the wood from being exposed to pathogenic fungi while also preventing evaporation.

Lathyrus spp.

Lathyrus spp.
Sweet Peas
FABACEAE

Zones: 5–9

The 150 species in this genus come from temperate regions in Europe, Asia, North and South America and the mountains of East Africa. They are grown for their colourful, usually fragrant flowers, attractive leaves, and vining habit. Newer cultivars are often scentless – if you want their wonderful perfume, read catalogue descriptions carefully to be sure you're buying a fragrant type. Fragrant cultivars of the annual *L. odoratus* include 'Old Spice', 'Royal Mixed' and 'Cuthbertson'. Perennial species are not always fragrant, either, but are excellent foils for trellises and screens that hide an unattractive part of the garden. *L. vernus* produces purple-blue flowers, and *L. sylvestris* produces pink flowers.

PROPAGATION METHODS

Easiest: *Seed.* These seeds must be scarified. Use fine sandpaper to lightly scratch their seedcoats, then soak them in warm water for about 8 hours. Plant under the soil surface in peat pots or soil blocks – they don't like to have their roots disturbed during transplanting. Keep them at 13° to 16°C while they are germinating. They germinate best in the dark, so you can put them in a closet or the basement if temperatures there are appropriate. They will germinate in 3 weeks to 1 month.

Additional methods: *Division.* Divide perennial sweet peas in early spring, before they have begun to grow vigorously.

Potential Problems

These plants are extremely easy to grow. If you treat the seeds as suggested but they don't germinate, suspect that you abraded the seedcoats too roughly – you just need to remove a bit of the outer skin so that they can absorb moisture. These plants are tough, so you shouldn't have much trouble dividing them as long as you replant them in an area with fertile, well-drained soils and keep them moist enough while they are becoming established.

Lavandula spp.
Lavenders
LAMIACEAE

Zones: 3–9

The 25 species in this genus come from the Mediterranean, North Africa, Asia and India. They are grown for their lovely, fresh fragrance; tall spikes of flowers; and ornamental leaves and habit. The flowers range in colour from pink to purple. Most species are cold-hardy; some, such as French lavender (*L. stoechas*), are grown as annuals in colder regions. 'Lavender Lady' will bloom before frost if you start it less than about 6 to 0 weeks before your last frost. Hardiness is sometimes a problem in Zone 5; if you have trouble keeping lavender alive over the winter months, grow *L. angustifolia* 'Hidcote' because it is especially hardy.

PROPAGATION METHODS

Easiest: *Seed.* Stratify seeds for 2 months and then start them inside, 6 to 8 weeks before the frost-free date. Keep their seed trays at 13° to 18°C. They will germinate in 2 weeks to 1 month or so, depending on species.

Additional methods: *Layering.* Many perennial species and cultivars don't reliably come true from seed, so you'll have to layer them if you want the same type of lavender plants. Layer in mid-spring; new plants will be ready to transplant in autumn or the next spring.

Semiripe cuttings. Take cuttings in late summer or early autumn and root them in a propagating chamber inside. You'll have to hold rooted plants inside over the winter because they'll be too tender to withstand outside weather.

Potential Problems

Seedlings are prone to damping-off diseases. Encourage them to grow as quickly as possible once they have germinated by placing them in bright indirect light and maintaining moist but not soggy conditions for their roots. Layering is a fairly foolproof way to propagate anything, including lavender. The only problem you might have is finding stems that bend easily to the soil surface. If this is the case, place heavy pots filled with soil around the plant and layer into these containers. Cuttings are always susceptible to fungal problems, but you can minimize these by providing sterilized containers and clean, fresh soil mix.

Lewisia spp.
Lewisias, Bitter Roots
PORTULACACEAE

Zones: 4–8

The 20 species in this genus come from western North America. They are grown for their good looks and tolerance to rocky soils. Evergreen species, such as *L. cotyledon*, thrive in partial shade or filtered sun, while deciduous species, such as *L. rediviva*, thrive in sun. Flowers range from clusters of cup-shaped blooms nestled in the leaves (*L. brachycalyx*) to solitary, single blooms held well above the foliage (*L.* 'George Henley').

PROPAGATION METHODS

Easiest: *Seed.* Plant seeds of any of these species, both evergreen and deciduous, after it is ripe in autumn. Place the seed tray in a cold frame to overwinter. This seed needs light to germinate; scatter it thinly on the surface of the medium, and mist it into niches. Alternatively, refrigerate ripe seed until you plant it in spring. Seed can take up to a year to germinate and does not do well if the temperature of the medium rises above 16°C.

Additional methods: *Offsets.* Evergreen species produce offsets that can be removed and replanted in midsummer.

Potential Problems

Forgetting about the seed is always possible when you have to wait a year or so for it to germinate. Place it where this will be impossible, and remember to protect it from overheating. Offsets must have roots to survive; divide them from their parent plant carefully, cutting so that you're taking roots as well as top growth.

Ligularia spp.
Leopard Plants, Golden Groundsels
ASTERACEAE

Zones: 4–8

Most of the 150 species in this genus come from Asia, but a few are from Europe. The plants are grown for their tall spikes or groups of blooms that are held well above the large, ornamental foliage. No matter what their form, all flowers are yellow or orange-yellow. Leaves are large, deeply cut or rounded, and often glossy, so these plants are useful in beds and borders even when they aren't in bloom. Cultivars of golden groundsel (*L. dentata*), such as 'Desdemona' and 'Othello', are among the most popular of these plants.

PROPAGATION METHODS

Easiest: *Seed.* Start seed early indoors, about 8 weeks before the frost-free date. Place where the soil mix will remain at temperatures of 13° to 16°C. The seed will germinate in 2 to 6 weeks.

Additional methods: *Division.* Divide plants in early spring, just before they start vigorous growth for the season.

Basal stem cuttings. Take cuttings in early summer, while the stem is still in vigorous growth but after it has reached about 13cm in length. Root it in a propagating chamber held at about 24°C.

Potential Problems

Seeds must remain moist, but not soggy wet, while germinating, so be sure to use a fast-draining medium. Plants divide easily. Replant in an environment that suits them and you won't have any trouble. Cuttings are always susceptible to fungal attack, but you can avoid problems with good sanitation and an appropriate environment.

Lilium spp.
Lilies
LILIACEAE

Zones: 3–10

The 100 species in this genus come from Europe, Asia and North America. They are grown for their showy, sometimes fragrant blooms. Flowers are white, yellow, orange, pink, red and purple and are enormously varied, ranging from trumpet shaped to bowl shaped to recurved and, finally, to funnel shaped. Popular garden species include the fragrant Oriental hybrids, the unscented but gorgeous Asiatic hybrids and the fragrant trumpet and aurelian hybrids. Whatever your tastes and garden needs, you'll be able to find a lily to satisfy them.

PROPAGATION METHODS

Easiest: *Division.* Lilies grow from bulbs that produce offsets or bulblets that can be divided when the foliage dies back. If they are very small, move them to a nursery bed for protection while they grow to blooming size.

Additional methods: *Division.* In addition to bulblets and offsets, you can divide scales from bulbs and propagate these (see page 72).

Some species, such as tiger lilies (*L. lancifolium*), produce bulbils in the leaf axils. Collect them when they are ripe by bagging the entire flower stalk with horticultural fleece or an old pillowcase when the flower petals drop, securing it tightly at the bottom. Bulbils will drop into the covering when they are ripe. Plant them in a tray, and place it in a cold frame over the winter. Bulbils will sprout in spring.

Seed. Save seed from your favorite species and even your cultivars, if you don't mind gambling on what the seeds will produce. Plant it as soon as it is ripe, and place the trays in the cold frame over the winter. Alternatively, you can hold it in the refrigerator over the winter months, and then plant it early inside. The seed requires light and temperatures of 16° to 24°C, so you may need to put the flat on a heating mat.

Potential Problems

Lilies are one of the easiest of all garden plants to grow. They can suffer from fungal infections in poorly drained soil but don't if the environment is appropriate. You won't have any trouble with offsets, bulblets or bulbils unless you plant them too deeply – remember that these are babies, and plant them only an inch or so below the soil surface. Keep them mulched over the winter. Seeds germinate easily. All you need is the patience to tend these plants for the 3 to 4 years it will take them to reach blooming size.

Limonium spp.
Statice, Sea Lavenders
PLUMBAGINACEAE

Zones: 3–9

The 150 species in this genus come from all over the world. They are grown for their lovely flowers, many of which are used in dry bouquets. Species range from annual to biennial to perennial. Flower colours include shades of white, yellow, orange, apricot, pink, purple and blue. New cultivars are being developed every year because this plant is becoming ever more popular. Look for *L. latifolium*, a perennial form with wide-spreading branches coated with small blue flowers; *L. tetragonum*, a biennial with pink flowers with white calyces; and cultivars of *L. sinuatum*, an annual form with flowers in every shade, many of which are bicoloured.

PROPAGATION METHODS

Easiest: *Seed.* This seed requires darkness to germinate; after seeding it, cover the container with plastic and then layers of cardboard or newspaper to exclude light. Maintain soil temperatures of 18° to 21°C. Seed will germinate in a little over 1 week to 3 weeks, depending on species.

Additional methods: Perennial forms can be divided in early spring. Replant immediately.

Potential Problems

It's rare to have difficulties with statice. The plants are remarkably disease resistant, so expect good results if you maintain an appropriate environment for them.

Lonicera spp.
Honeysuckles
CAPRIFOLIACEAE

Zones: 3—9

The 180 species in this genus come from all over the northern hemisphere. They are grown for their sweetly fragrant flowers that coat the branches in spring and summer. Some climb and some are shrubs; some are evergreen and others are deciduous. Flowers may be bicoloured and are white, pink, yellow, light orange and ivory. If you want a woody climber, consider growing *L.* x *americana* with its fragrant yellow and reddish flowers or *L.* x *brownii* with its red flowers with yellow centres. Among the deciduous shrubs, *L.* x *purpusii* 'Winter Beauty' and *L. xylosteum* 'Emerald Mound' are good choices. Both have attractive blooms, and *L.* x *purpusii* smells delicious in late winter and early spring.

PROPAGATION METHODS

Easiest: *Layering.* Climbing species can easily be layered. Set the branches in place in early spring. They will be ready to transplant by autumn or the following spring.

Additional methods: *Softwood cuttings.* Take cuttings in spring, when they are 10 to 15cm long, and root them in a soil-less mix.

Semiripe cuttings. Take cuttings in mid- to late summer, and root them in a propagating chamber with mist.

Potential Problems

Honeysuckle is easy to propagate. Softwood cuttings root so easily that you can even root them in water, as long as you change it daily to prevent fungal diseases from occurring.

Lupinus spp.
Lupines
FABACEAE

Zones: 4–8

The 200 species in this genus come from the Mediterranean; North Africa; and North, Central and South America. They are grown for their showy flower spikes and large plush leaves. Because they are legumes, they are an excellent soil conditioner, too, adding nitrogen as well as a great deal of biomass. Flowers are often bicoloured, in shades of blue, white, red, yellow, pink and purple. If you are looking for good garden specimens, try growing one of the Russell hybrids; they come in every colour and have a nice architectural form.

PROPAGATION METHODS
Easiest: *Seed.* You'll need to scarify or soak the seed. If you do both, soak the seed for no more than the time it takes to plump up; otherwise, soak it for 24 hours. Seed requires darkness to germinate, so cover the seeding container with a thick layer of newsprint. Maintain soil temperatures of 13° to 16°C. Seeds will germinate sporadically in the next 2 weeks to 2 months.

Additional methods: *Basal cuttings.* Take cuttings from new growth arising from the crown when it is about 15cm long. Cuttings will lose too much water via transpiration if you leave them intact. Instead, remove at least half of the leaf before you place the cutting in the rooting medium. Keep root cuttings at a temperature of 16° to 18°C.

Potential Problems
Seeds should never pose any problems. Lupines germinate so easily that they self-seed in good conditions. These cuttings can be tricky. They are really susceptible to fungal infections, so it's important to maintain strict cleanliness in the area where they are rooting; keep air circulation high; and use only fresh, clean media and sterilized containers.

Malus spp.
Apples, Crabapples
ROSACEAE

Zones: 3–9

The 35 species in this genus come from Europe, Asia and North America. These plants are grown primarily for their fruit or, in cases where the fruit is not appetizing, clusters of fragrant spring flowers. Blooms are in shades of white and pink, and the prominent stamens are a bright yellow. Regional adaptations are extreme enough that it's wise to buy plants only from area nurseries. Some species and cultivars are self-infertile, so you'll need a compatible pollinator if you want fruit. Again, check with your local suppliers to learn good combinations for your area.

PROPAGATION METHODS
Easiest: *Grafting.* Whip-and-tongue and bud grafting are the most frequently used for apples and crabapples. Graft in late winter or very early spring – just before the sap begins to run. Bud graft in late summer.

Additional methods: *Seed.* Species can be propagated with seeds. Plant ripe seeds in autumn and place their container in a cold frame over the winter. They will germinate sporadically over the spring and early summer months.

Potential Problems
Follow grafting directions carefully to keep the wood from being exposed to fungal infections or drying out before the union forms. Seeds are reliable. If you want to grow a particular rootstock, check speciality seed companies to learn if seed is available.

Monarda spp.
Bee Balms, Bergamot
LAMIACEAE

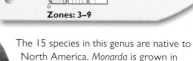
Zones: 3–9

The 15 species in this genus are native to North America. *Monarda* is grown in contemporary gardens for its long-lasting, tall spikes of red, pink and light purple flowers that grow in tiers. Native Americans used it as a tea, and colonists

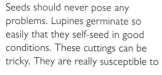

Lonicera spp.

adopted this use. It's also a wonderful bee flower, so farmers often grow it near hives and plants that need insect pollination. Old favourites include 'Cambridge Scarlet', 'Blue Stocking' and 'Croftway Pink'.

PROPAGATION METHODS

Easiest: *Seed.* Plant seeds inside about 8 weeks before your frost-free date in spring. Keep the starting medium about 16° to 21°C during the day, but move it off the heating mat at night so that the temperature falls to 10° to 13°C. Seeds will germinate in 2 weeks or so.

Additional methods: *Division.* Divide in spring, before vigorous growth begins. Plants develop a "dead spot" in the centre of the clump after about 3 years, so you really need to divide them and discard the dead section every few years.

Softwood cuttings. Take cuttings in early summer, and root them at a temperature of about 24°C.

Potential Problems

You're unlikely to have any trouble with growing *monarda* from seed. Similarly, division is foolproof as long as you take a minimum of care. Cuttings are always susceptible to fungal infections, but these plants root easily enough so that you are unlikely to have trouble as long as you root them in a fresh, clean medium and sterilized starting container.

Nandina domestica
Heavenly Bamboo
BERBERIDACEAE

Zones: 6–9

This species from India, China and Japan is grown for its lovely plant habit, foliage, flowers and fruit. Flowers appear in mid-summer and are small, white, star-shaped florets with prominent yellow anthers that cover 36- to 41cm-long conical panicles. The fruit is bright red. Cultivars have been bred to be tiny: 'Wood's Dwarf' is only 46cm tall, in comparison to the species, which is 1.8 m tall, and 'Nana', which can rise to 1.2 m.

PROPAGATION METHODS

Easiest: *Semiripe cuttings.* Take cuttings when they have begun to harden up at the base, and root in a misting propagating chamber or on a soil heating mat held at 24°C. Set a pebble tray next to the rooting container, and tent both containers with plastic film.

Additional methods: *Seed.* Seeds are available from speciality seed companies, or you can collect your own. Plant them as soon as they are ripe in autumn, and place the starting containers in a cold frame to overwinter. They will germinate in spring.

Potential Problems

Cuttings root easily if the environment is correct. Don't let them get cool, and remember to introduce fresh air to their tent a couple of times a day. Seeds are variable, but if you use the whole packet, you'll get enough plants to start a small grove. Unlike bamboo, its namesake, *Nandina*, does not sucker, and you'll have to plant a number of seedlings to create a small grove or screen.

Nelumbo spp.
Lotus
NELUMBONACEAE

Zones: 4–11

The two species in this genus come from Asia, Australia and North America. They are grown for their large, circular leaves that rise above the surface of shallow water and their showy, fragrant flowers with prominent centres. The flowers of American lotus (*N. lutea*) are a soft, pale yellow colour, while cultivars of sacred lotus (*N. nucifera*) are white, pink, lavender, red and bicoloured.

PROPAGATION METHODS

Easiest: *Seed.* Scarify seed before planting it in spring, and then soak it in 38°C water until it plumps up. Plant it in a humus-rich loam in small, individual pots, and cover the loam with about 5cm of water. Keep the containers at a temperature of 25° to 27°C until the seeds germinate. Keep transplanting the lotus to larger containers and increasing the depth of the water covering the roots until they are large enough to transplant to the margins of your pool.

Additional methods: *Division.* Underground rhizomes will form as the lotus grows. Divide these in spring, replanting them immediately.

Potential Problems

Seeds germinate readily and do not suffer from damping-off diseases. If you can keep the temperatures high enough, but not too high, you won't have trouble. Division can be tricky. The plants resent disturbance, so work carefully and quickly. Have the new pots filled and standing next to where the plant has been growing. Make a fast, decisive cut. Replant the divisions immediately, always keeping them wet. Replace at the correct depth in the water.

Nymphaea spp.
Waterlilies
NYMPHAEACEAE

Zones: 4–11

The 50 species in this genus come from all over the world. They are grown for their showy flowers and large, decorative leaves that float on the water. Flowers may be single or double and are white, yellow, pink and blue. Many species and cultivars are fragrant, but some have no scent at all. When purchasing a water lily, check to see if it is hardy in your region; some are frost sensitive, while others survive under the ice in a pond.

PROPAGATION METHODS

Easiest: *Division.* Divide rhizomes or separate offsets of plants in spring. Some species produce plantlets. Divide these in summer, pot them up individually, and, as you do with small offsets, place their pots in shallow water.

Additional methods: *Seed.* If collecting your own, enclose the seed pod in horticultural fleece soon after the petals fall. The seed pod will sink under the water level and release seeds when the seeds are ripe. When this happens, plant them in loamy, humus-rich soil. Place the pots where they are covered with about 2cm of water. Maintain temperatures of about 10° to 13°C for species that are winter-hardy and 21° to 27°C for tropical species.

Potential Problems

Divisions will establish themselves easily if they are in an appropriate soil mix and at the proper depth in the pool. However, don't submerge small offsets or plantlets so deeply that their leaves can't float. Seeds can float away in rough water. If you are concerned about this, cover the top of the pot with wide-mesh netting. When the first stem emerges, it will find its way to an opening in the mesh.

Onoclea sensibilis
Sensitive Fern
DRYOPTERIDACEAE

Zones: 4–9

This plant comes from eastern Asia and eastern North America. It grows wild in damp, sheltered sites with dappled shade and fertile, humus-rich, acid soil. In gardens, it's used for its ornamental value in damp, somewhat shady areas where other plants may have difficulty growing. The leaves of this plant die down at the first frost, leaving the dried remains of the spore-bearing fronds standing stiffly until snows either bury them or knock them over.

PROPAGATION METHODS
Easiest: *Division*. These plants grow from rhizomes that elongate with age. Left to its own devices, a small patch of sensitive ferns will slowly enlarge to eventually cover the entire area where conditions are favourable to its growth. You can lift the rhizomes and divide them in early spring. Replant immediately.

Additional methods: *Spores*. Propagating from spores usually requires tight environmental control, but you may have success without the need for specialized equipment or painstaking care if you simply sprinkle the spores on bare, humus-rich, acid soil in a tray when they are ripe. Don't cover the spores with soil. Place the tray where temperatures will remain at about 16°C, light is diffused and humidity is very high. Spores take several weeks to several months to germinate.

Potential Problems
You should have no trouble with divisions as long as you replant them in appropriate environments. Spores are tricky – so much so that they are not covered in this book. However, it's possible that spores of a sensitive fern will germinate without special treatment, so it's worth a try if you're intrigued by the challenge.

Origanum spp.
Oreganos, Marjorams
LAMIACEAE

Zones: 4–9

The 20 species in this genus come from the Mediterranean and southwest Asia. Many species, such as oregano (*O. vulgare*) and marjoram (*O. majorana*), are grown to use in the kitchen or in fragrant dried and fresh flower and herb arrangements. Smaller plants are frequently grown in rock gardens or between the stones on a path because they are tough and resilient. If growing oregano for culinary purposes, choose only the white-flowered forms. However, for all other purposes, the lovely pink flowers are a standout in the late summer garden.

PROPAGATION METHODS
Easiest: *Seed*. Start seed early in spring, usually about 8 weeks before the last frost date. Sprinkle seed on the soil surface of the tray and mist it into niches because it germinates better with light. Maintain soil temperatures of 13° to 16°C. Germination takes 1 week; 2 at most.

Additional methods: *Division*. Many species of oregano grow from creeping rhizomes. Divide these in spring. You can also divide plants that grow from a crown in spring. In each case, replant immediately.

Potential Problems
Problems are rare with this plant unless the environment is consistently humid and the air is still. In that case, leaf-spot diseases or stem rots can develop. If this has been a problem in your garden, replant divisions or seedlings where they will experience slight breezes.

Oxalis spp.
Shamrocks, Sorrels
OXALIDACEAE

Zones: 5–10

The 500 species in this genus come from Africa and South America and are incredibly diverse. Some are fibrous rooted; others have rhizomes, bulbs or tubers. The genus includes annuals and perennials. In gardens in temperate climates, they are generally grown for the ornamental value of their leaves and

Oxalis spp.

the cup-shaped flowers. Plants such as *O. regnellii* 'Atropurpurea' grow well in partial shade, but others, such as *O. tetraphylla* 'Iron Cross', thrive in sun, although they will tolerate a bit of filtered shade in the afternoon.

PROPAGATION METHODS

Easiest: *Seed.* Plant seeds early inside; generally, about 8 to 10 weeks before the frost-free date. Keep the soil medium at about 16° to 21°C.

Additional methods: *Division.* Divide tubers, rhizomes, bulblets or fibrous-rooted plants in early spring, before they begin vigorous growth.

Potential Problems

Problems are unlikely. These plants are sturdy; so as long as you fulfill their environmental needs, which vary by species, you should have no trouble propagating them.

P

Pachysandra spp.

Pachysandras, Spurges
BUXACEAE

Zones: 4–9

The four species in this genus come from China, Japan and the southeastern United States. This plant is grown for ground cover, partially because of its attractive leaves and partially because it tolerates shade as long as the soil is moist but not soggy. *P. terminalis* 'Variegata' is one of the most popular cultivars because of its white or pale yellow variegated leaves and because, out of all cultivars of this species, it grows the slowest.

PROPAGATION METHODS

Easiest: *Division.* Japanese spurge (*P. terminalis*) plants develop long stolons that root at the nodes. Simply snip these off in spring or fall, and replant them in a spot with moist soil and at least partial shade. They will take root.

Additional methods: *Seed.* Some species produce seed, but it's impractical to grow it from seed unless you are working to create your own hybrids. In that case, bag the inconspicuous flowers after the petals fall so you can collect the seed as it drops. Sprinkle it on the soil surface of a seed tray, and place

the tray in a cold frame to overwinter. Seed will germinate in spring.

Potential Problems

The most common difficulty with Japanese spurge is keeping it in bounds. This plant takes over as much ground as it can, so you are unlikely to need to do much more than cut off pieces of the stolon to propagate it. Allegheny spurge (*P. procumbens*) is a clump-forming species and will not spread.

Paeonia spp.

Peonies
PAEONIACEAE

Zones: 3–8

The more than 30 species in this genus come from Europe, Asia and North America. They are grown for their large, showy flowers; glossy, decorative leaves; and impressive plant habit. Flowers are single, semidouble, double and "anemone form" – a single or double bloom with clusters of tiny, pointed petals in the centre, in place of stamens. Plants may grow into bushy shrubs or small trees with drooping branches. It's hard to choose between the hundreds of stunning cultivars; once you develop a taste for peonies, you're likely to want to collect them.

PROPAGATION METHODS

Easiest: *Division.* Some people have a knack with peonies and have no trouble dividing them; others do not. But even if you fall into the latter category, practice may make perfect. Divide peonies in early autumn and replant so that the eyes are only 5cm below the soil surface. Mulch them after the soil freezes, and remove the mulch after the last spring frost.

Additional methods: *Seed.* Seeds are easy only if you have patience. Plant them in seed trays in autumn, when they are ripe, and set them in a cold frame for the winter. Some may germinate the following spring and summer, others will wait for the following year, and still others will wait until the third year after they have been planted. Keep the tray from overheating or drying during this long wait.

Semiripe cuttings. Take cuttings from tree peonies in summer, when they are ready. Root them in a propagating chamber.

Root cuttings. Plant them in pots, and sink the pots under a cold frame for winter.

Potential Problems

Peonies are not easy to propagate. Their roots do not like disturbance, so it's important to move quickly and take as much of the soil surrounding the division as possible. Seeds will

eventually germinate, but you have to care for their seed trays a long time. Cuttings are always vulnerable to fungi. Use fresh, clean media and sterilized pots, and maintain a clean environment in the area where they are rooting.

Papaver spp.

Poppies
PAPAVERACEAE

Zones: 2–8

The 70 species in this genus come from Europe, Asia, South Africa, Australia, North America and subarctic regions. This diverse group of plants is grown for the vividly coloured blooms. Species range from annual to biennial to perennial. Flowers are single or double, and colours include shades of white, yellow, orange, pink and red. The distinctive seed pods are often used in dried arrangements. Favourite garden species include cultivars of Iceland poppies (*P. croceum*), including 'Champagne Bubbles' and 'Wonderland'; cultivars of Oriental poppies (*P. orientale*), including 'Cedar Hill' and 'Checkers'; and cultivars of corn or Shirley poppies (*P. rhoeas*), including 'Mother of Pearl' and 'Picotee Mixed'.

PROPAGATION METHODS

Easiest: *Seed.* Poppies germinate easily and also self-seed, so you may need to thin and transplant your volunteers every spring rather than start new seedlings. But to get started, plant seeds where they are to grow in the garden just before the frost-free date or 6 weeks earlier inside in peat pots or soil blocks. Their roots resent disturbance, so they are tricky to transplant if you have to remove them from a starting container. With the exception of Oriental poppies, which need light for germination, you'll need to cover the containers with a thick layer of newspaper to exclude light. Hold the containers at a temperature of 13°C. Seeds will germinate in 1 to 2 weeks.

Additional methods: *Root cuttings.* Take root cuttings from perennial species in late autumn to early winter, and hold them in sunken peat pots surrounded by plastic in a cold frame over the winter. They will develop into plants in the spring and early summer.

Potential Problems

Do not let poppy seeds overheat; if they do so, they will go dormant or die. Protect young seedlings from fungal diseases by keeping air circulation high. Root cuttings are fairly reliable as long as you don't bury them too deeply in the soil or plant them upside down. Be careful not to disturb the roots when you are taking off the plastic around the peat pots before planting the poppies in their permanent garden positions.

Penstemon spp.
Beard Tongues
SCROPHULARIACEAE

Zones: 3–9

The approximately 250 species in this genus come from North and Central America. They are grown for their lavish production of tall spikes of flowers from midsummer to autumn. Colours include shades of white, pink, purple, red and yellow. Species vary in size and preferences, from small rock garden plants such as creeping penstemon (*P. caespitosus*) and pineleaf penstemon (*P. pinifolius*) to tall border plants such as *P. digitalis* 'Husker Red' and *Penstemon* 'Pennington Gem'.

PROPAGATION METHODS

Easiest: *Seed*. Plant seeds inside about 8 to 10 weeks before the frost-free date. This seed must be stratified, so chill it for at least a month in the refrigerator before planting it. Seeds require light and soil temperatures of 13° to 16°C to germinate. In good conditions, they will germinate within 3 weeks at the most.

Additional methods: *Division*. Divide in early spring, before vigorous growth begins.

Softwood cuttings. Take cuttings in early summer, when they are approximately 15cm long.

Semiripe cuttings. Take cuttings in mid- to late summer, and root them in a propagating chamber if possible.

Potential Problems

Seeds germinate easily. Many penstemon grow best in well-drained soils. Seedlings of these plants are vulnerable to fungal root rots if their starting mix is too water retentive. Add sharp sand or perlite to the mix if it seems to hold too much water. Cuttings are always susceptible to fungal infections. Root softwood cuttings at a temperature of about 18° to 24°C to speed up the rooting process and minimize the time when they are most vulnerable. Remember to introduce fresh air into the propagating chamber a few times a day when rooting semiripe cuttings.

Pereskia spp.
Rose cacti, Barbados gooseberries
CACTACEAE

Zones: 9–10

The 16 species in this genus come from Florida, Mexico, Central America, South America and the West Indies. In most areas, they are grown for their ornamental leaves and showy flowers and are often used as a large, container-grown houseplant in areas north of Zone 9. In Central America and the West Indies, the edible leaves are used in cooked dishes. Flowers are creamy white, pink or purple, with prominent golden centres. The rose cactus (*P. grandifolia*) is an evergreen plant that flowers from spring to autumn.

PROPAGATION METHODS

Easiest: *Seed*. Plant seed in the spring, and maintain soil temperatures of 16° to 24°C in the starting flat. Seeds will germinate within 3 weeks.

Additional methods: *Softwood cuttings*. Take 15cm-long cuttings from new growth in very early spring, before the plant blooms.

Semiripe cuttings. If you live where you can grow this plant outside, it will develop what appears to be woody growth. Take cuttings from the current season's growth when it is starting to harden at the base. Root in a propagating chamber kept at 24°C.

Potential Problems

Seeds are reliable and won't cause problems if you plant them shallowly in a quickly draining medium and do not overwater them at any time. Cuttings are susceptible to fungal infections but root so quickly that you are unlikely to have trouble with them.

Phalaenopsis spp.
Moth Orchids
ORCHIDACEAE

Zones: 9–10

The 50 species in this genus come from Asia and Australia. They are grown for their arching racemes of long-lasting, beautiful flowers and ornamental leaves. This is one of the easiest orchids to grow in the average home because it tolerates a fairly wide range of relative humidity levels. Nonetheless, you'll have to mist it a few times a day in a dry environment if you want it to thrive. Favourite *Phalaenopsis*

hybrids include the bicoloured pale and rose-pink Doris, the snowy white Allegria, and the white and deep pink Yukimai.

PROPAGATION METHODS

Easiest: *Division*. Don't try to divide the root system of this plant – it's more than likely to die if you do. However, many species and cultivars produce offsets at their bases that can be divided from the parent plant once they have a strong enough root system to survive on their own.

Additional methods: *Plantlets*. Other species produce plantlets on the flower spikes. Bend these down and pin them to the soil surface of a pot filled with orchid medium. Once they have successfully rooted and are growing new leaves, you can sever them from the parent plant.

Softwood cuttings. Very few people are successful with this method of propagation, but professional growers routinely increase their stock this way. Take cuttings as soon as they are about 15cm long, and root them in a propagating chamber.

Potential Problems

If you wait to divide offsets or plantlets until roots are well developed, you should have no problems with them. Cuttings may rot before they root, but if the environment in your propagating chamber is just right and the rooting medium is extremely fast draining, you may have success.

Philadelphus spp.
Mock Oranges
HYDRANGEACEAE

Zones: 5–8

The 40 species in this genus come from Europe, Asia and North and South America. They are grown for their fragrant white flowers that coat the branches in late spring as well as their glossy leaves and shrubby plant habit. Flowers may be single, as in 'Beauclerk', or double, as in 'Buckley's Quill' and 'Dame Blanche'.

PROPAGATION METHODS

Easiest: *Softwood cuttings*. Take cuttings as soon as terminal shoots are 10 to 15cm long. Root them at temperatures of 21° to 24°C.

Additional methods: *Semiripe cuttings*. Take cuttings as soon as their bases have begun to harden up, generally in mid- to late summer.

Hardwood cuttings. Take cuttings in early winter, and treat them as appropriate for your climate. They will root the following spring.

Potential Problems

Mock orange cuttings root easily. If at first you don't succeed with softwood cuttings, try again with semiripe cuttings, and if they don't work, try hardwood cuttings. You'll probably succeed with all three methods and end up with so many plants, you have to give them to everyone you know.

Phyllostachys spp.

Bamboos
POACEAE

Zones: 6–10

The 80 species in this genus come from Asia and the Himalayas. They are grown for their straight and variously coloured stems, which range from the gold of the fishpole bamboo (*P. aurea*) and *P. aureosulcata* var. *aureocaulis* to the green of most species to the dark stems of black bamboo (*P. nigra*). Leaves vary as well, with both 'Albovariegata' and 'Holochrysa' having white stripes.

PROPAGATION METHODS

Easiest: *Division*. This plant grows from rhizomes. Running bamboos spread vigorously and can threaten to take over your garden, especially in warm areas. No matter where you live, you will never have any difficulty propagating it. Simply dig and divide the rhizomes.

Additional methods: None necessary.

Potential Problems

This plant reproduces itself so freely that your only problem will be keeping it in bounds. Sink a 30cm-deep piece of aluminium sheeting around the area in which you want it to stay. You may have to thin new plants that grow up within this zone, but very few will get beyond it.

Physostegia spp.

Obedient Plants
LAMIACEAE

Zones: 4–8

The 12 species in this genus come from eastern North America. They are grown for their spikes of flowers that look equally stunning in an outside border and in a vase. Their common name comes from an odd characteristic – if you bend the flower stem, it will remain in that position. This, in combination with the long vase life, adds to its appeal as a cut flower. Flowers are white or various shades of pink, rose, and light purple, and favorite cultivars include 'Alba', 'Bouquet Rose' and 'Variegata', with its white variegated leaves.

PROPAGATION METHODS

Easiest: *Seed*. Seeds require stratification in the refrigerator for at least 2 but preferably 6 weeks. Plant them in a mix containing humus, and place them in a position where the medium will maintain temperatures of 13° to 24°C during the day, but move them where the medium temperature will be 7° to 10°C at night. They will germinate within 3 to 4 weeks.

Additional methods: *Division*. Divide plants in early spring, before they begin new growth.

Potential Problems

Problems are rare with these plants. They are easy to start if you take care that their medium is at the correct temperature both day and night and doesn't dry out. Similarly, divisions establish easily as long as you don't disturb the roots when they are in active growth.

Pieris spp.

Pieris
ERICACEAE

Zones: 5–9

The seven species in this genus come from Asia, the Himalayas, North America and the West Indies. These lovely shrubs are grown for both their ornamental foliage and their clustered panicles of small flowers. Leaves are whorled and, in some cultivars, variegated with an edging of white on every leaf. 'Forest Flame' has leaves that begin as a dark, glossy red. They then turn pink, pale even more to become ivory, and finally become green. Flowers are stunning, too. They are white or pink and liberally coat the branches of the bush.

PROPAGATION METHODS

Easiest: *Seed*. Plant the seed in autumn and hold it in the cold frame over winter, or store it inside and wait until spring to plant it. It does not require stratification to germinate. Researchers report that even though it doesn't need light to germinate, the rate of germination is increased if the seed tray is exposed to light. Keep it at 24°C. Seed will germinate within 1 to 2 months.

Additional methods: *Softwood cuttings*. These are notoriously difficult to root, but it is possible. Place them in a propagating chamber held at 24°C.

Potential Problems

Seeds have a very high germination rate. Collect them in the autumn, when the pods are dry, and separate the empty seedcoats from those that are viable. As long as their starting area provides an appropriate environment, they will germinate. But they are very slow growing and are susceptible to fungal infections while they are young. It's not unusual for a month to go by between the time they produce their first and second leaves. Keep air circulation high and temperatures around 16° to 24°C during this time.

Pinus spp.

Pines
PINACEAE

Zones: 3–10

The 120 species in this genus come from all over the world. They are grown as specimen plants because of their robust good looks; as windbreaks and hedgerow plants because they are dense; and, in the case of dwarf cultivars, as edging plants or accents in a mixed border. When choosing a pine for your garden, take the time to do some research; you could discover that you can grow a Scots pine (*P. sylvestris*) cultivar such as 'Aurea' or 'Gold Coin' that has bright yellow needles or a white pine (*P. strobus*) cultivar such as 'Contorta' with its long, twisted, blue-green needles on contorted branches.

PROPAGATION METHODS

Easiest: *Seed*. Seed for many pines takes 2 years to mature in the cone. But when the cones fall, the seed is likely to scatter to the ground. To collect it, either bag cones within your reach as the scales begin to open or lay old sheets under the tree when cones begin to fall; check daily for seeds. Plant the seeds in autumn in trays that you set under a cold frame. They require stratification and will receive it naturally in this position. In spring, remove them from the cold frame and place them in a protected position in the garden or against a wall of the house. They will germinate over the next few months. If you want to be sure to get every last seedling, overwinter the seed tray under the cold frame for a second year.

Additional methods: *Grafting*. Spliced side-veneer grafting is appropriate for pines.

Potential Problems

Be careful not to let the seed trays overheat under the cold frame glazing. If you think this is a possibility, cover the cold frame with light-obscuring material so it doesn't heat up on sunny winter days. Seeds germinate easily but grow slowly. Make certain that their medium drains well and that they are in a position with good air circulation while you are waiting for them to be large enough to transplant to an outdoor nursery bed.

Platycodon grandiflorus
Balloon Flowers
CAMPANULACE

Zones: 4–9

This species comes from Asia. It is grown for its stunning flowers and ability to thrive equally well in full sun and partial shade. Flowers are white, blue, violet, pink, or lilac. They are called balloon flowers because their buds expand to create large globes before opening into five-petalled blooms. They hold well in the vase, so they make dramatic cut flowers.

PROPAGATION METHODS

Easiest: *Seed.* Plant seed early inside; because they can be slow growing, it's best to start them about 8 to 10 weeks before the frost-free date. They require light to germinate and fare best when their starting medium is about 21°C. They will germinate in 2 to 4 weeks.

Additional methods: *Division.* Some species produce basal shoots that form roots. Check for these before you decide to lift a plant and divide it. These plants are very slow to emerge in spring – so much so that you may despair that they died over winter. But don't go digging them up to explore; chances are, they are just taking their time to emerge. Wait until new growth shows before digging and dividing them. Replant immediately.

Potential Problems

Seeds are unlikely to cause any problems as long as you keep air circulation high and temperatures moderate in the area where the seedlings are growing. In general, these plants have a very high germination rate, so it's reasonable to expect nothing but success. Balloon flowers also divide easily. If you replant the divisions in good soil that is well drained, they should be fine.

Primula **spp.**
Primroses
PRIMULACEAE

Zones: 4–8

Almost half of the 425 species in this genus come from the Himalayas, with the balance being widely distributed; most come from the northern hemisphere, but a few species come from the southern hemisphere. Primroses are grown for their spring flowers; various species are suitable for rock gardens, others for bog plantings, and many for mixed beds and borders in full sun, partial shade, or even deep shade. No matter what your conditions, you can probably find a primrose to suit it. Of the many cultivars available, favourites include the Gold Laced group, *P. japonica* 'Postford White' and the Dreamer series.

PROPAGATION METHODS

Easiest: *Seed.* Start seed of tender species in spring and that of hardy species either in autumn, in trays destined for a cold frame, or in spring, about 8 weeks before the frost-free date. In either case, seed of hardy species must be stratified in the refrigerator for at least 3 weeks. All seeds except *P. sinensis* require light to germinate. Maintain soil temperatures of 13° to 16°C for all species.

Additional methods: *Division.* Some species produce offsets that can be divided from the parent plant when they have a root system that's adequate to support them. Other species can be lifted and divided in early spring.

Root cuttings. Take root cuttings in early winter, after the plants are dormant. They will develop roots and shoots in spring.

Potential Problems

Seeds are generally reliable. However, they are susceptible to fungal infections in humid conditions with stagnant air. Make certain that the medium drains well, and keep air circulation good once the seeds have germinated. Divided plants will

Pinus spp.

re-establish easily if you have replanted them in appropriate conditions. Expect a high proportion of your root cuttings to develop into plants. But protect them from overheating over the winter, and don't let the seed tray sit under glazing in the bright spring sun or you'll bake the roots. Be patient – it may take all spring and early summer for all of the cuttings to develop.

Prunus spp.

Peaches, Nectarines, Plums, Cherries
ROSACEAE

Zones: 4–8

The more than 200 species in this genus come from North America, Europe, Asia and South America. They are grown for their fruit or – more commonly in small city and suburban gardens – their decorative flowers and naturally graceful, vaselike forms. The bark of species such as *P. maackii* and *P. serrula* is so stunning that it serves as a focal point in the winter garden, while the dark red-purple leaves of *P. virginiana* 'Shubert' and *P. cerasifera* 'Nigra' make them stand out during the growing season. Flowers are white or pink and form in clusters that coat the branches in spring. Finally, even if you don't want to eat the fruit, grow a tree such as bird cherry (*P. avium*) for the birds in your neighbourhood.

PROPAGATION METHODS

Easiest: *Grafting.* These trees respond very well to whip-and-tongue grafting but are likely to form unions if you choose another technique.

Additional methods: *Seed.* Species can be started from seed. Let the fruit rot around the seeds, and then plant the seeds in a seed tray that you leave under a cold frame for the winter. Cover the cold frame to protect it from overheating.

Greenwood cuttings. Take greenwood cuttings as soon as they are 10 to 15cm long.

Softwood cuttings. Take softwood cuttings as soon as they are ready, generally in mid- to late spring.

Potential Problems

Grafts must be taped securely so that no air enters and the cambium layers of the scion and rootstock remain in contact with each other. Seeds will germinate sporadically; be patient, and keep the tray well watered and protected from excessive rainfall or heat. Cuttings are always susceptible to fungal attack, but both greenwood and softwood cuttings of *Prunus* species root quickly enough

that you will have a high success rate as long as you start them in fresh, clean media; maintain temperatures of 21° to 24°C; and don't let the air around them stagnate.

Quercus spp.

Oaks
FAGACEAE

Zones: 4–9

The 600 species in this genus come from all over the northern hemisphere. Today they are grown for their striking good looks, although their acorns have been used as a nut by many cultures. Gardeners in northern areas are likely to recall a deeply lobed, glossy green leaf when they think of oak trees. However, leaves are far more varied, especially among the southern species. Laurel oak (*Q. laurifolia*) has leaves that look like those of a bay laurel tree, as does the live oak (*Q. virginiana*). In contrast, the leaves of the California live oak (*Q. agrifolia*) resemble those of hollies, with small spines on the margins.

PROPAGATION METHODS

Easiest: *Seed.* Acorns sprout easily. Collect them as they drop, plant them in a deep pot, and hold the pot in a somewhat protected spot or under a vented cold frame in winter. They will germinate in the spring.

Additional methods: *Grafting.* Use a whip-and-tongue or spliced side graft on deciduous oaks. Evergreens respond best to spliced side-veneer grafts; use rootstock plants that are

3 or 4 years old, and don't cut off the top of the rootstock plant until the graft has been in place a full year.

Semiripe cuttings. Take cuttings from vigorous new growth, just as the base is hardening up. Root the cuttings in a propagating chamber, if possible, to speed their root development.

Potential Problems

Acorns have enormously long first roots, so if you plant them in a shallow container, they will immediately start curling around themselves. Avoid this by planting them in a pot that is at least 30cm deep. Sink the pot in the soil over the winter to give it some protection. Grafts fail if they dry out while the union is forming or the cambium layers don't touch. Wrap any graft well to avoid problems. Cuttings are always susceptible to fungal diseases, but if you use fresh, clean media and sterilized rooting containers and remember to introduce fresh air to their propagating area, you should have no trouble with these plants.

Rhododendron spp.

Rhododendrons, Azaleas
ERICACEAE

Zones: 4–10

The more than 800 species in this genus come from Europe, Australia, Asia, North America, India and New Guinea. They are grown in gardens for their flowers and bushy plant habit. Many are evergreen, with large, leathery, dull or glossy green leaves.

Quercus spp.

Depending on species and cultivar, flower colours include white, yellow, pink, orange, salmon, red, magenta and purple. The stamens of most species are long and prominent, making the bloom look light and airy. There are so many cultivars and they are so regionally adapted that it's difficult to pick out favourites. However, if you are looking for a fragrant, late-blooming azalea, try *R. arborescens*. For a fragrant rhododendron, try 'Fragrantissimum'; for one that is covered with pale pink-salmon buds that open to clusters of soft yellow flowers, you'll want 'Golden Torch'.

PROPAGATION METHODS

Easiest: *Layering*. If you can bend a branch down to the soil, try layering it in autumn. These plants root easily, so you may have a plant that is transplant size by the following year.

Additional methods. *Seed*. Use seed only for species, not for cultivars. These plants hybridize easily, so it's completely possible to create your own cultivar. Collect the seed as soon as it is ripe, and plant it in a very acid soil mix. Place it in a cold frame to overwinter. Alternatively, collect the seed and stratify it in the refrigerator until spring. Keep the planting medium at temperatures of 13° to 18°C while the seed is germinating.

Softwood cuttings; semiripe cuttings. Both types are appropriate. Take the cuttings as soon as they are ready, and root them in a propagating chamber if possible. If not, keep humidity levels high while they are rooting.

Grafting. Graft plants in late winter. Use a saddle or a side-wedge graft.

Potential Problems

Layering is fairly foolproof if the branch you've chosen is young enough. Older branches do not layer well. Seeds can take 12 to 18 months to germinate; be patient, and keep their starting pot moist and at the correct temperature during this time. In warm climates, you may need to store it in the basement to maintain appropriate temperatures. Check on it frequently; it would be a shame to lose a plant simply because you hadn't noticed that it had already germinated. Protect cuttings from fungal diseases by using fresh, clean media and sterilized containers. Grafts can be difficult, so choose another method of propagation if possible. If not, be certain that the cambium layers touch and the wood doesn't dry out.

Rosa spp.
Roses
ROSACEAE

Zones: 2–10

The 150 species in this genus come from Asia, Europe, North Africa and North America. They are grown for their flowers as well as their various plant habits. This group is so diverse that no matter where you live or what you want from a rose, you'll be able to find it. The new 'English Roses', many of them bred by David Austin, are excellent for all gardeners, especially beginners. They are bred from a combination of various old garden roses and modern roses with qualities such as disease resistance and the ability to rebloom or bloom a little for the whole season. Look for cultivars such as 'Jayne Austin', 'Constance Spry' and 'Gertrude Jekyll'.

PROPAGATION METHODS

Easiest: *Budding*. Graft buds in summer (see page 128).

Additional methods: *Softwood cuttings*. Take these as soon as they are 10 to 15cm long. Root them in a misting propagating chamber for best results.

Hardwood cuttings. Take these in late autumn and overwinter them outside, buried in medium under a cold frame or in a protected spot in the garden.

Layering. Layer stems in spring, and leave them in place for a year.

Seed. Seed of species roses will germinate quite easily, which is handy if you need to grow a rootstock for some other favourite roses or you want to create a large planting of a rose such as *R. rugosa*. Plant the seed as soon as it is ripe in a pot that you place in a cold frame over the winter or in a protected spot in the garden. Seeds will germinate sporadically over the next year or so.

Potential Problems

Budding is a fairly easy technique for the manually dexterous. If you wrap the bud and surrounding wood well to exclude air, you shouldn't have a problem. All cuttings are susceptible to fungal attack, but if you root them in clean media and maintain good environmental conditions, they should root before they are attacked. Layering works with some roses but not all. Try it and find out how your roses respond. Be patient with seeds, and maintain their starting containers well. Be certain that they don't dry out or overheat.

Rosa spp.

Rosmarinus spp.
Rosemaries
LAMIACEAE

Zones: 7–10

The two species in this genus come from the Mediterranean. They are grown for their good looks as well as their use as a culinary herb. Most species and cultivars of rosemary are hardy only to protected spots in Zone 8. However, a relatively new introduction, 'Arp', is reliably hardy to Zone 7. Gardeners in more northern areas often plant their rosemary bushes in containers so they can move them inside and outside, according to the season. 'Golden Rain' is a cultivar with foliage edged in bright yellow when it first emerges, and 'Prostratus' is a prostrate form that makes a nice display in a hanging basket.

PROPAGATION METHODS

Easiest: *Seed.* Seed is widely available for a number of cultivars of rosemary. Plant it in early spring, and maintain temperatures of about 21° to 24°C in the starting medium. It will germinate within 2 weeks to 1 month.

Additional methods: *Layering.* Layer plants just after they bloom in summer. Keep the layers in place until the following spring, when they should be large enough to sever from the parent plant. But look before you cut – if they don't yet have a good root system, leave them in place until autumn.

Semiripe cuttings. Take cuttings in summer, when they are ready. Root them in a high-humidity environment, such as a misting propagating chamber, or mist them frequently during the day while they are rooting.

Potential Problems

Seedlings grow slowly at first, so they are quite close to the medium. If it is too moist, they run the risk of damping off. Be certain that the medium is fast draining, and don't overwater it. Layers take easily, so you shouldn't have a problem with them. Cuttings are always vulnerable to fungal infection. Keep the humidity high where they are rooting, but don't let the air stagnate.

Saintpaulia spp.
African Violets
GESNERIACEAE

Zones: 10–11

The 20 species in this genus come from East Africa. They are grown as flowering houseplants in the rest of the world because outdoor temperatures are often too hot or too cold—they prefer living at the same temperatures we do. There are literally thousands of cultivars, primarily of *S. ionantha.* The most important to know when you are propagating them are the "chimera" types— their petals are striped, as in the popular cultivar 'Concord'.

PROPAGATION METHODS

Easiest: *Seed.* If you are growing a species rather than a cultivar, plant seeds in spring. Plant them as soon as they are ripe, and place them where the temperature of their starting medium can range between 18° to 24°C.

Additional methods: *Division.* Many plants form suckers at the base, and a few form plantlets on their flower stalks. Divide and replant these as soon as they have a large enough root system so they can take in water and nutrients while becoming established.

Leaf petiole cuttings. For all cultivars but chimeras, this is an excellent method. Place them in a misting propagation chamber, and maintain temperatures of 24° to 27°C while they are rooting. Chimeras will not come true from leaf cuttings but will from suckers or plantlets.

Potential Problems

Fungal diseases are the bane of African violets. Maintain the suggested temperatures while seeds are germinating and cuttings are rooting. Divisions are not as susceptible to difficulties as seeds or cuttings. For cuttings, introduce fresh air to the propagating chamber to minimize problems.

Salix spp.
Willows
SALICACEAE

Zones: 2–9

The 300 or so species in this genus come from all over the world. They are grown for their graceful habits – many of them are weeping – as well as the colourful bark of some species. This was once a common medicinal plant because it contains the same active ingredient as aspirin; the remedy for toothache or headache was to chew on a twig of a willow tree. Among the many species and cultivars, a few truly stand out. For example, one cultivar of the weeping willow (*S. babylonica* var. *pekinensis* 'Tortuosa') grows twisted branches that can be an effective focal point. Cultivars of the white willow (*S. alba*) include *S. alba* var. *vitellina*, with shoots that are bright yellow in winter, and *S. alba* var. *vitellina* 'Britzensis', with red-orange shoots.

PROPAGATION METHODS

Easiest: *Softwood cuttings.* These plants root so easily that you may discover roots on the bottoms of stems you've brought in to use in a cut flower display. But instead of relying on that, take softwood cuttings as soon as they are 10 to 15cm long, and root them in fresh, clean media.

Additional methods: *Hardwood cuttings.* Take these in early winter, and hold them in a container in the cold frame until spring.

Grafting. Use a whip-and-tongue graft on these trees.

Potential Problems

It's rare to have any difficulties with willow cuttings – they root incredibly easily as long as their environmental needs are met. Keep hardwood cuttings from overheating in the cold frames during the winter and make sure that air circulation is high around softwood cuttings, and you'll never have a problem. Grafts also take easily as long as the cambium layers are touching and the wood is not allowed to dry out.

Sansevieria spp.
Mother-in-Law's Tongues, Snake Plants
AGAVACEAE

Zones: 9–11

The 60 species in this genus come from Africa, Madagascar, India and Indonesia. North of Zone 9, they are generally grown as houseplants because of their variegated leaves and plant habit. Some cultivars of *S. trifasciata*, such as 'Laurentii', grow as tall as 1.2 m high, while others, such as 'Golden Hahnii' and 'Hoops Pride', are smaller and form spreading rosettes. Left to their own devices, they will spread to fill a container, and in warm environments, they form spikes of small green flowers. The blooms aren't attractive, but they are very fragrant and make viable seeds.

PROPAGATION METHODS
Easiest: *Division.* Plants produce suckers at the base. Leave these in place until they have a large enough root system to survive on their own, and then divide them.

Additional methods: *Leaf cuttings.* Take these cuttings whenever you can provide a warm, humid environment for them to root.

Potential Problems
If you divide suckers too soon, the plants will suffer. Check to be sure that you are taking roots when you take the suckers. These cuttings root easily, so you are unlikely to have difficulties. However, the plant does not like soggy soils, so keep the medium moist but not wringing wet to minimize the chances of a fungal attack.

Sedum spp.
Stonecrop
CRASSULACEAE

Zones: 4–11

The approximately 400 species in this genus primarily come from the northern hemisphere, but a few come from South America. They are grown for their interesting forms and ability to withstand dry conditions. Of the many cultivars available, 'Autumn Joy' is one of the most popular; the plant is loved for its deep red, late-summer blooms that gradually change to mahogany brown in autumn and continue to stand until buried by snow. Burro's tail (*S. morganianum*) is a favorite hanging-basket plant that survives well in any household.

PROPAGATION METHODS
Easiest: *Division.* Divide perennials in spring.

Additional methods: *Seed.* Plant annuals and biennials in early spring. Place them where the temperature of their medium will range around 13° to 16°C. Plant tender species in spring as well, but keep their seeding medium a bit warmer – 16° to 18°C. If you have seed of a hardy cultivar or species, plant it in a seed tray in autumn, and set the tray under a cold frame or in a protected spot in the garden for winter.

Softwood cuttings. Take cuttings of vegetative, rather than flowering, growth in early summer. Let the cut end form a callus and then place it in the rooting medium. Root them in a warm environment.

Leaf petiole cuttings. Many species root well from a single leaf. Again, let the wound callus before placing the leaf in the rooting medium.

Potential Problems
Divisions will reestablish well if you keep their soil moist but not soggy while they are rooting. If the soil is soggy, they run the risk of developing root rot. See that the wound has dried and a small callus has formed before placing a cutting in the rooting medium. Keep the cuttings at temperatures around 18° to 21°C while they are rooting.

Solenostemon spp.
Coleus
LAMIACEAE

Zones: 10–11

The 60 species in this genus come from tropical areas of Africa and Asia. They are grown for their ornate leaves of green, yellow, salmon, red and inky purple. Leaves are often variegated, with picoteed margins or veins of a different colour than the body. They are well-loved houseplants but are becoming more popular as summer bedding plants in areas where they are not hardy. 'Black Prince' and 'Black Dragon' are new cultivars with colouration so deep it truly looks black, and 'Laser Red' is a true red. Spend some time searching for new cultivars if you are looking for a foliage plant to perk up a spot in the mixed border.

PROPAGATION METHODS
Easiest: *Seed.* Start seeds inside about 8 weeks before the frost-free date. They require light for germination, so scatter the seed thinly on vermiculite spread over a fast-draining soil mix, and mist the seeds into niches. Cover the container with plastic wrap, and set it where the medium will maintain temperatures of about 24°C.

Additional methods: *Softwood cuttings.* Take cuttings of fresh, terminal growth at any time of year. Root them in a medium with a temperature of about 24°C. They will root very quickly.

Potential Problems
Problems with these plants are rare. They grow easily from seed and are so easy to root that they sometimes do it in water. Resist the urge to start them this way because "water roots" have a hard time adjusting to the texture of soil. Start them in a soil-less medium.

Stephanotis spp.
Bridal Wreaths, Waxflowers
ASCLEPIADACEAE

Zones: 9–11

The 13 species in this genus come from Madagascar, China, Japan and Cuba. The plants are grown for their vining habit and intensely fragrant white flowers that coat the stems in summer. *S. floribunda* is commonly grown as a houseplant in northern areas. Although you can grow this plant in a hanging basket, its natural habit is to twine around nearby structures or trees and grow upward; plants grown on a trellis will thrive.

PROPAGATION METHODS
Easiest: *Seed.* Maintain temperatures of 24° to 27°C during the period while seeds are germinating.

Additional methods: *Softwood cuttings.* Take tips of stems when new growth is 10 to 15cm long. Root cuttings in a propagating chamber if possible. If not, tent the area, and add a pebble tray to keep relative humidity levels high.

Layering. Layer stems just before they begin vigorous new growth for the season.

Semiripe cuttings. Take these in midsummer, when the base of the new wood has begun to harden. Again, use a propagating chamber if possible.

Potential Problems
Seeds germinate sporadically, so it's important to take care of the seed tray for at least 3 months before deciding that no more seeds are going to germinate. Layers are unlikely to suffer problems as long as you leave them in place long enough that they develop a good root system. Cuttings are always vulnerable to fungal attack, so keep the area scrupulously clean – remember to clean and disinfect the pebble tray frequently.

Syringa spp.
Lilacs
OLEACEAE

Zones: 3–8

The 20 species in this genus come from Europe and Asia. They are grown for their fragrant flowers that bloom in early spring as well as their graceful habit. Flower colours include white, pink, lavender, blue, purple and a deep pink-red. Favourite species and hybrids include *S. vulgaris*, the common lilac, with its many cultivars; Chinese lilac (*S. x chinensis*); and Persian lilac (*S. x persica*). Look for cultivars such as *S. patula* 'Miss Kim' if you want a traditional lilac-coloured bloom, *S. vulgaris* 'Sensation' if you want a lavender flower edged in white, and *S. vulgaris* 'Leon Gambetta' if you want a deep purple flower.

PROPAGATION METHODS
Easiest: *Suckers*. *S. vulgaris* and its cultivars sucker readily and will gradually enlarge to many times the initial width. Divide suckers that have a large enough root system to sustain growth on their own.

Additional methods: *Layering*. Layer plants in early summer, and sever the layers from the parent plant in the autumn, if they have a large root system, or the following spring.

Greenwood cuttings. Take cuttings in spring, as soon as they are about 10 to 15cm long. Root with bottom heat, and mist if possible. If not, tent the area and place a pebble tray in it to increase relative humidity.

Seed. Species will come true from seed. Plant ripe seed in trays, and hold the trays under a cold frame to overwinter. They will germinate sporadically in spring.

Potential Problems
Suckers die if you sever them from the parent plant before they are ready to live on their own. Always dig down to check on the size of the root system before cutting off the plant. Patience repays you. Layers must also have a good root system; don't move them too early. Cuttings are always vulnerable to fungal attack, but lilac cuttings root easily and quickly, so you are unlikely to have problems. Seeds can take some time to germinate – keep the tray for a year if no seeds germinate the first spring.

Tolmiea menziesii
Pick-a-Back Plants, Youth on Ages
SAXIFRAGACEAE

Zones: 6–9

This genus comes from western North America, where it grows as an understory plant in coniferous forests. In the garden, it is grown as ground cover under shade-giving plants or as an oddity because of its unusual method of propagating itself. The cultivar 'Taff's Gold' is prized for the golden variegation on the leaves. The plant is a close relative of Coral Bells (*Heuchera* spp.), and the leaves have that same thin, delicate feeling.

PROPAGATION METHODS
Easiest: *Division*. Remove the plantlets that grow on the top of the leaves, just where the petiole connects to the leaf, or, even easier, peg a leaf with a developing plantlet to the soil surface on a pot containing a fast-draining medium and let the plantlet develop roots into the soil. The mother leaf will naturally decompose, and you can cut the new plant from the parent plant.

Additional methods: *Division*. This plant forms creeping rhizomes that can be divided in early spring.

Seeds. Seeds are available from specialty companies. They require stratification in the refrigerator for at least 1 month. Germinate at temperatures of 18° to 24°C. The seeds germinate readily.

Potential Problems
You should have no problems propagating this plant as long as you do not expose it to bright light. Remember that it is an understory plant from woods with filtered light, and give it the same sort of environment.

Verbascum spp.
Mulleins, Ornamental Mulleins
SCROPHULARIACEAE

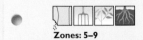

Zones: 5–9

The approximately 360 species in this genus come from Europe, North Africa and Asia. In gardens, most species are grown for their ornamental value. Their tall spikes of showy flowers rise above the mound of leaves that retain their beauty through spring, summer and autumn. Among the many available species and cultivars, standouts include 'Cotswold Queen', with yellow blooms with prominent brown eyes; *V. chaixii* f. *album*, with white blooms with orange-red eyes; and 'Pink Domino', with rose-coloured flowers.

Syringa spp.

PROPAGATION METHODS

Easiest: *Seed.* Seeds require at least a month of stratification. After that, you can plant them inside, 6 to 8 weeks before the frost-free date. Place their seed trays where the soil medium will retain temperatures of 13° to 16°C. Germination will occur in 2 weeks to 1 month.

Additional methods: *Division.* Divide perennials in spring, before growth resumes.

Semiripe cuttings. Take cuttings when the base of the new growth has begun to harden, and root in a propagating chamber if possible. If not, place on bottom heat, and mist several times a day.

Root cuttings. Hold these cuttings under a cold frame over the winter. They will develop top growth following spring and summer.

Potential Problems

Seeds are reliable. If the environment is appropriate for the plants, you shouldn't have any trouble with them. Divisions are pretty well foolproof as long as you take large enough divisions with both top growth and roots. Cuttings are always vulnerable to fungal attack. Make certain that the environment is clean when you are rooting semiripe cuttings. Root cuttings will rot if their medium is allowed to be soggy – make sure it drains well.

Veronica spp.
Speedwells, Veronicas
SCROPHULARIACEAE

Zones: 3–8

The 250 species in this genus come mainly from Europe, where they grow in areas as diverse as moist swamplands and rocky, sunny hills. Not surprisingly, you can find a Veronica that will suit whatever environment you have. This genus includes annuals as well as perennials, and most species produce flower spikes above the leaves in mid- to late summer. Flower colours are frequently in shades of blue, although there are also white, purple, red and pink types. *V. spicata* is one of the most common species and includes hundreds of cultivars, such as 'Alba' and 'Minuet'. 'Sunny Border Blue' is a long-blooming hybrid.

PROPAGATION METHODS

Easiest: *Seed.* This seed requires a month of stratification as well as exposure to light. Plant it early inside, about 6 to 8 weeks before the frost-free date, and set it where the soil will retain temperatures of about 21°C during the day. But take it off the heating mat in the evening – it requires temperatures of 10° to 16°C at night. Germination occurs in 2 weeks to 1 month.

Additional methods: *Division.* Divide perennial species in early spring or autumn.

Softwood cuttings. Take softwood cuttings as soon as they are about 10 to 15cm long.

Potential Problems

Seed is reliable as long as it is stratified and sees light while it is germinating. Place it on the surface of a seed tray filled with a fast-draining medium covered with a thin layer of vermiculite. If you take large enough divisions, you should have no trouble dividing this plant. Softwood cuttings root quickly. If you root them in a fresh, clean medium and keep their area clean, you can expect success.

Viburnum spp.
Viburnums
CAPRIFOLIACEAE

Zones: 4–9

The approximately 150 species in this genus come from northern temperate areas all over the world, although a few come from as far south as South America and southeast Asia. The plants are grown for their bushy habit, often fragrant flowers, bird-feeding berries and gorgeous autumn foliage. Blooms are in shades of white and pink. Of the many available species and cultivars, favourites include Koreanspice viburnum (*V. carlesii*), *V. nudum* 'Winterthur' and Chinese snowball (*V. macrocephalum*).

PROPAGATION METHODS

Easiest: *Seed.* Although it may seem strange to start such a large plant from seed, this is a reliable method of propagating Viburnum species. Stratify ripe seeds in the refrigerator all winter long, or plant them and place their starting tray under a cold frame for the winter. No matter where you stratify them, set the starting trays outside, under protection from intense light and heat and protected from rain. They will germinate sporadically over the spring and summer months.

Additional methods: *Greenwood cuttings.* Take these cuttings from deciduous species as soon as they are ready in spring. Root them over bottom heat, and mist them several times each day.

Semiripe cuttings. These cuttings are appropriate for evergreen species. Take them in mid- to late summer, when the base of new growth has begun to harden, and root them in a propagating chamber if possible. If not, use bottom heat set to 24°C, and mist them several times during the day.

Potential Problems

Patience pays with seeds – don't give up on a tray of seeds for at least a year. Remember to keep it protected from temperature extremes year round, not just in summer, and maintain soil moisture at moderate levels. Viburnum roots well from cuttings, so you can expect success if you keep the area clean, use fresh rooting media and introduce fresh air into the propagating chamber.

Wisteria spp.
Wisterias
FABACEAE

Zones: 5–9

The 10 species in this genus come from both Asia and the United States. Plants are grown for their striking appearance as well as their drooping racemes of intensely fragrant flowers. The vines are large and strong – they require a sturdy trellis. Blooms are white or shades of blue, violet or pink. Chinese wisteria (*W. sinensis*), Japanese wisteria (*W. floribunda*) and American wisteria (*W. frutescens*) all boast numerous cultivars. Favourites include *W. floribunda* 'Black Dragon' and *W. sinensis* 'Alba' and 'Prolific'.

PROPAGATION METHODS

Easiest: *Layering.* Layer in autumn, and leave the layer in place until the following year, when it should have a well-developed root system and be able to survive on its own.

Additional methods: *Softwood cuttings.* Take cuttings from the new growth coming from the crown in spring, as soon as shoots are 10 to 15cm long.

Root cuttings. Take cuttings in winter, and root them over bottom heat.

Potential Problems

Layering is a reliable method of propagating wisteria and should give you no problems, as long as you wait to lift the new plant until it has an adequate root system. Cuttings are always vulnerable to fungal attack, but wisteria is an easy plant to root. If you observe commonsense sanitation guidelines, you shouldn't have any trouble.

Glossary

A

Annual: A plant that completes its life cycle in one season.

Anther: The part of the flower on which pollen forms.

Apical dominance: The tendency of the topmost shoot on a plant to inhibit the development of lateral buds growing lower on the plant.

Air layering: A propagation method in which a stem is stimulated to develop roots while still on the plant.

Asexual reproduction: Any sort of reproduction other than by seeds. Common techniques include cuttings, divisions, layers and grafts.

Auxin: A hormone-like substance in plants that controls growth.

B

Basal: Referring to growth at the bottom of a plant.

Basal plate: The bottom of a bulb. New bulbs, or bulblets, can grow from the basal plate.

Biennial: Referring to a plant that completes its life cycle in two seasons.

Budding: Grafting using a bud of the scion rather than a part of the branch.

Budding knife: A specialized knife that gardeners use for taking buds for grafting as well as opening the area on the rootstock where the bud will be inserted.

Bud stick: The branch from which buds are taken for grafting.

Bulb: A fleshy underground structure from which plants such as tulips, lilies and daffodils grow.

C

Callus: A protective layer of tissue that forms over wounds.

Cambium: A layer of tissue with the capacity to form new cells in stems as well as roots.

Chip budding: Grafting a small section of scion wood that contains a bud to a rootstock.

Cold frame: A small outside structure meant to give plants some protection from outside environmental factors such as extreme cold, drying winds and pounding rain.

Compatible: In propagation, referring to rootstocks and scions that can be grafted to each other or the male and female parts of flowers that can join to produce a seed.

Corm: A fleshy storage structure from which a new plant will grow. In contrast to bulbs, corms are solid structures.

Crown: The top of the rootstock at or just below the soil line where shoots begin their growth.

Cultivar: A variety of a plant that has been developed through breeding.

Cutting: A section of a plant that is used for propagation. Cuttings can be taken from roots or stems at various stages of growth.

D

Dibble: A pointed tool used to make holes in soil where plants can be inserted.

Dicot: A plant that produces two seed leaves.

Dioecious: Referring to a plant species that produces male and female flowers on separate plants.

Dropping: A propagation technique in which a plant is dropped, or lowered, into the soil to stimulate rooting from the branches.

E

Eye: A bud that has not yet developed into a shoot.

F

Fertilization: The union of male and female reproductive parts.

G

Germination: The process in which a seed sprouts and the first root and stem begin to grow.

Grafting: Attaching one part of a plant, usually a portion of a branch or bud, to another plant so that they will form a union and grow as one plant.

Grafting tape: Specialized tape that is somewhat elastic and sticks to itself.

Greenwood cutting: A cutting taken from a stem that is in a quickly growing, vegetative state.

H

Hardwood cutting: A cutting taken from a stem when the bark is mature and the wood has fully hardened for the winter.

Herbaceous plant: A plant with soft, non-woody stems. In cold climates, the top growth of herbaceous plants generally dies down over winter.

Humidity: The amount of moisture in the air.

Hybrid: A plant that was produced by crossing parents of different species.

I

Internode: The area between two nodes on a stem.

Interstem: In grafting, a section of a stem of a species different from the rootstock and the scion that is grafted between the two in order to confer characteristics such as dwarfing.

Isolation: In seed-saving, the practise of growing plants of the same genus so far apart that they cannot breed or of erecting physical barriers that prevent pollen from other than the selected plants to fertilize the flowers.

L

Latent bud: A bud that is not developing but that has the capacity to develop if conditions change.

Lateral bud: A bud on the side rather than the bottom or top of a branch.

Layering: A method of propagation in which a stem is stimulated to develop roots. Types of layering include: tip layering, in which the tip of the branch is used; simple layering, in which only one section of the branch is stimulated to root; serpentine layering, in which several sections of the branch are stimulated to root; and French or trench layering, in which many nodes along the branch are stimulated to root.

Leaching: In seed starting, the practise of running water over seeds long or frequently enough to wash off any chemicals that naturally inhibit germination.

Leaf cutting: A portion of a leaf that is used for propagation.

M

Mallet cutting: A stem cutting that includes a mallet-shaped portion of the previous year's growth.

Mature wood: Wood that has hardened and developed fully formed bark.

Meristem: Area of a plant with cells capable of developing into leaves, flowers, stems or roots.

Monocot: A plant that produces only one seed leaf.

Monoecious: Referring to a plant species that produces separate male and female flowers on the same plant.

Mound layering: A type of layering in which earth is heaped over the crown and bottom of the stems of a plant.

Mutation: A change in the genetic code of a cell. Mutations happen naturally as a consequence of environmental effects.

N

Node: An area on a stem, branch or root where new branches or buds develop.

O

Offset: A new plant that grows from the crown or a stolon of a parent plant.

Open pollinated: The term used to describe a plant that will remain true as long as it is fertilized with pollen from the same species.

Ovary: The part of a flower that contains the eggs that will develop into seeds if they are fertilized by pollen from a compatible plant.

P

Parent plant: A plant that is used to produce other plants.

Perfect flower: A flower that contains both male and female parts.

Petiole: A leaf stalk.

pH: The measure of the acidity/alkalinity ratio.

Pollen: The male reproductive particles of a flower that confer genetic information to the new plant.

Pollination: The transfer of pollen from a male flower to a female flower. Pollination can be carried out by insects, water, wind or the gardener's hand.

Propagation unit: An enclosure where the environment, often including relative humidity as well as temperature, can be regulated so that seeds or cuttings can have the best possible conditions in which to develop.

R

Rhizome: A fleshy storage structure from which roots and stems can grow.

Root cutting: A portion of a root used to stimulate the growth of a new plant.

Rooting compound: A chemical substance, in either liquid or powder form, that stimulates root growth.

Rootstock: The plant that is used as the roots for a grafted plant.

Runner: A common name for plant stolons that run along the surface of the soil or just under it. New plants can grow from nodes along the length of the runner.

S

Scale: In lilies, a section of the bulb that is capable of developing into a fully formed bulb and subsequent plant.

Scarification: The process of nicking or scratching the seed coat so that water can more easily penetrate it.

Scion: The plant that is used as the top growth of a grafted plant.

Self-incompatible: A flower or plant that is unable to pollinate itself.

Semiripe cutting: A cutting taken from wood that has begun to mature.

Softwood cutting: A cutting taken from a stem when it is in its rapid first growth.

Spore: A reproductive unit, generally of a fern or moss, from which a new plant can develop.

Stigma: The part of a flower that leads to the ovary. The stigma usually has a somewhat sticky top and a tube down which pollen can fall.

Stratification: The process of cooling or freezing seeds in order to break dormancy. Some seeds require alternate stratification, or alternate warming and cooling.

Sucker: A shoot that grows from the roots or crown of a plant.

T

Transpiration: The plant process of releasing water vapor through the stomates, or pores, in the leaves and stems.

Tuber: An enlarged area of an underground stem that includes eyes, or dormant buds, from which new plants can grow.

U

Union: The area where the rootstock and a scion have grown together.

V

Viability: The capacity of a seed to germinate and develop. Seeds of each species tend to be viable for a certain number of years.

W

Widger: A tool used to lift seedlings.

Winter annual: A plant that lives only one year, generally germinating in midsummer and forming seeds and dying the following spring.

Woody plant: A plant that develops woody tissue.

Hardiness zones

Remember that hardiness is not just a question of minimum temperatures. A plant's ability to survive certain temperatures is affected by many factors, such as the amount of shelter given and its position within your garden.

Europe

KEY

Average annual minimum temperature

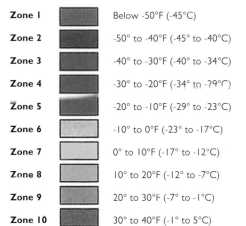

Zone 1		Below -50°F (-45°C)
Zone 2		-50° to -40°F (-45° to -40°C)
Zone 3		-40° to -30°F (-40° to -34°C)
Zone 4		-30° to -20°F (-34° to -29°C)
Zone 5		-20° to -10°F (-29° to -23°C)
Zone 6		-10° to 0°F (-23° to -17°C)
Zone 7		0° to 10°F (-17° to -12°C)
Zone 8		10° to 20°F (-12° to -7°C)
Zone 9		20° to 30°F (-7° to -1°C)
Zone 10		30° to 40°F (-1° to 5°C)

South Africa

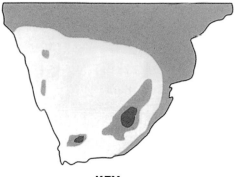

KEY

Average annual minimum temperature

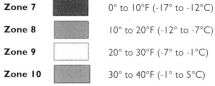

Zone 7		0° to 10°F (-17° to -12°C)
Zone 8		10° to 20°F (-12° to -7°C)
Zone 9		20° to 30°F (-7° to -1°C)
Zone 10		30° to 40°F (-1° to 5°C)

Australia and New Zealand

Plant list

ABELIA spp.: hardwood cuttings
ABELIOPHYLLUM *distichum*: semi-ripe cuttings
ABELMOSCHUS *moschatus*: seed
ABIES:
   spp.: seed, grafting, hardwood cuttings
   A. concolor: seed, grafting, hardwood cuttings
   A. nordmanniana: seed, grafting, hardwood cuttings
   A. veitchii: seed, grafting, hardwood cuttings
   A. vejarii: seed, grafting, hardwood cuttings
ABRONIA:
   A. pogonantha: seed
   A. villosa: seed
ABUTILON:
   spp.: seed, semiripe cuttings, softwood cuttings, greenwood cuttings
   A. pictum: seed, semiripe cuttings, softwood cuttings
ACACIA spp.: greenwood cuttings
ACANTHUS:
   spp.: seed, division, root cuttings
   A. hungaricus: seed, division, root cuttings
   A. mollis: seed, division, root cuttings
   A. spinosus: seed, division, root cuttings
ACER:
   spp.: seed, grafting, softwood cuttings
   A. griseum: seed, grafting, softwood cuttings
   A. japonicum: seed, grafting, softwood cuttings
   A. palmatum: seed, grafting, softwood cuttings
   A. pensylvanicum: seed, grafting, softwood cuttings
   A. rubrum: seed, grafting, softwood cuttings
   A. saccharinum: seed, grafting, softwood cuttings
ACHILLEA spp.: seed, division, softwood basal cuttings
ACHIMENES spp.: division
ACONITUM spp.: seed, division
ACTAEA spp.: division, root cuttings
ACTINIDIA spp.: semiripe cuttings, hardwood cuttings, layering
ADIANTUM:
   spp.: division, plantlets, spores

   A. caudatum: plantlets
   A. pedatum: division, spores
AECHMEA spp.: seed, offsets
AESCLEPIAS *tuberosa*: seed
AESCULUS *parviflorus*: suckers
AGAPANTHUS spp.: division
AGAVE spp.: seed, offsets
AGERATUM spp.: seed
AGLAONEMA:
   spp.: cane cuttings
   A. modestum: offsets
AJUGA:
   spp.: division, seed
   A. genevensis: division, seed
   A. pyramidalis: division, seed
   A. reptans: division, seed
ALCEA *rosea*: seed
ALCHEMILLA spp.: seed, division
ALISMA spp.: division, seed
ALLIUM:
   spp.: division, seed
   A. caeruleum: division
   A. cristophii: division, seed
   A. flavum: division, seed
   A. giganteum: division, seed
   A. schoenoprasum: division, seed
   A. tuberosum: division, seed
ALNUS *spp.*: greenwood cuttings
ALOCASIA spp.: division
ALOE:
   spp.: offsets, leaf cuttings, seed
   A. arborescens: offsets, leaf cuttings, seed
   A. aristata: offsets, leaf cuttings, seed
   A. haworthioides: offsets, leaf cuttings, seed
   A. mitriformis: offsets, leaf cuttings, seed
   A. variegata: offsets, leaf cuttings, seed
ALSTROEMERIA spp.: division
AMELANCHIER:
   spp.: suckers, layering, root cuttings, softwood cuttings
   A. x grandiflora: suckers, layering, root cuttings
   A. lamarckii: suckers, layering, root cuttings
   A. stolonifera: suckers, layering, root cuttings
AMMI *majus*: seed
ANANAS spp.: seed, offsets
ANEMONE:
   spp.: division, seed, root cuttings
   A. blanda: division, seed, root cuttings
   A. coronaria: division, seed,

   root cuttings
ANTHURIUM spp.: offsets
ANTIRRHINUM *majus*: seed
AQUILEGIA spp.: seed
ARALIA spp.: stooling
ARISTOLOCHIA spp.: semiripe cuttings
ARMERIA spp.: seed, division
ARONIA spp.: suckers
ARTEMISIA:
   spp.: division, semiripe cuttings, seed
   A. abrotanum: seed, division, semiripe cuttings
   A. absinthium: division, semiripe cuttings, seed
   A. annua: seed, division, semiripe cuttings
   A. dracunculus: seed, division, semiripe cuttings
ARUM spp.: division
ARUNCUS spp.: division
ASARUM:
   spp.: division, seed
   A. canadensis: division, seed
   A. europaeum: division, seed
   A. hartwegii: division, seed
   A. shuttleworthii: division, seed
ASCLEPIAS *tuberosa*: seed
ASPARAGUS:
   spp.: division, seed
   A. densiflorus: division, seed
   A. officinalis: division, seed
   A. setaceus: division, seed
ASPERULA:
   spp.: seed, division
   A. odorata see **GALIUM** *odoratum*
   A. orientalis: seed, division
   A. suberosa: seed, division
ASTER spp.: division
ASTILBE:
   spp: division
   A. x arendsii: division, seed
   A. chinensis: division, seed
   A. japonica: division, seed
   A. simplicifolia: division, seed
   A. thunbergii: division, seed
AUCUBA *japonica*: softwood cuttings, semiripe cuttings, layering
BAPTISIA:
   B. australis: seed, division
   B. lactea: seed, division
   B. pendula: seed, division
   B. perfoliata: seed, division
BEGONIA:
   spp.: greenwood cuttings, division, seed

   B. bowerae: leaf vein cuttings, upright leaf cuttings
   B. x hiemalis: leaf vein cuttings, upright leaf cuttings
   B. masoniana: leaf vein cuttings, upright leaf cuttings
   B. rex-cultorum: leaf vein cuttings, upright leaf cuttings
   B. Semperflorens group: seed
   B. x tuberhybrida: division
BELAMCANDA:
   spp.: seed, division
   B. chinensis: seed, division
BERBERIS spp.: softwood cuttings, semiripe cuttings, mallet cuttings, seed
BERGENIA spp.: division, seed, root cuttings
BETULA spp.: greenwood cuttings, grafting
BILLBERGIA spp.: seed
BLECHNUM *spicant*: offsets
BOUGAINVILLEA spp.: layering, air layering
BRACTEANTHA *bracteatum*: seed
BROMELIA spp.: seed, offsets
BRUGMANSIA spp.: greenwood cuttings
BRUNFELSIA spp.: greenwood cuttings
BRUNNERA *macrophylla*: root cuttings
BUDDLEJA:
   spp.: seed, semiripe cuttings, hardwood cuttings
   B. alternifolia: seed, semiripe cuttings, hardwood cuttings
   B. davidii: greenwood cuttings, softwood cuttings
   B. globosa: seed, semiripe cuttings, hardwood cuttings
BUPLEURUM *fruticosum*: semiripe cuttings
BUXUS
   spp.: semiripe cuttings, division, seed, greenwood cuttings, softwood cuttings

CALADIUM spp.: division
CALANDRINIA spp.: seed
CALATHEA spp.: seed
CALENDULA *officinalis*: seed
CALLA *palustris*: division, seed,
CALLICARPA spp.: softwood cuttings, semiripe cuttings
CALLISTEMON spp.: greenwood cuttings
CALLUNA *vulgaris*: layering,

semiripe cuttings, dropping

CALOCEDRUS *decurrens:* semiripe
cuttings

CALYCANTHUS:
spp.: suckers, layering,
softwood cuttings, seed,
greenwood cuttings
C. floridus: softwood cuttings

CAMASSIA spp.: division

CAMELLIA spp.: semiripe cuttings,
hardwood cuttings, air
layering, grafting

CAMPANULA:
spp.: seed, division, softwood
cuttings
C. carpatica: seed, division,
softwood cuttings
C. persicifolia: seed, division,
softwood cuttings

CAMPSIS spp.: semiripe cuttings,
seed, hardwood cuttings, root
cuttings

CANNA:
spp.: division, seed
C. x *generalis:* seed

CARDAMINE *pratensis:* division

CARDIOCRINUM spp.: division

CARYA *illinoinensis:* grafting

CARYOPTERIS x *clandonensis:*
softwood cuttings

CASTANEA *mollissima:* grafting

CASTILLEJA *chromosa:* seed

CATANANCHE *caerulea:* root
cuttings

CATTLEYA spp. seed

CEDRUS spp.: hardwood cuttings

CELOSIA spp.: seed

CENTAUREA *cyanus:* seed

CEPHALARIA *gigantea:* root
cuttings

CEPHALOCEREUS spp.: seed,
grafting

CERATOSTIGMA spp.: semiripe
cuttings

CERCIS spp.: greenwood cuttings

CEREUS spp.: seed

CEROPEGIA spp.: division

CHAENOMELES:
spp.: seed, softwood cuttings,
hardwood cuttings, semi-
ripe cuttings
C. speciosa: greenwood
cuttings, softwood cuttings

CHAMAECYPARIS spp.: hardwood
cuttings, grafting

CHELONE spp.: division, seed,
softwood cuttings

CHIONODOXA spp.: division

CHLOROPHYTUM spp.: offsets

CHRYSANTHEMUM spp.: division,

seed, softwood cuttings

CIMICIFUGA see Actaea

CISSUS:
spp.: layering
C. antarctica: seed

CITRUS spp.: air layering, grafting

CLEMATIS:
spp.: layering, division,
softwood cuttings, seed
C. florida: layering, division,
softwood cuttings, seed
C. tangutica: layering, division,
softwood cuttings, seed

CLETHRA:
spp.: suckers, layering
C. alnifolia: softwood cuttings
C. arborea: suckers, seed,
semiripe cuttings
C. barbinervis: suckers, seed,
semiripe cuttings

CLIVIA spp.: offsets

CODIAEUM:
spp.: seed
C. variegatum: greenwood
cuttings, air layering,
stooling

COLCHICUM spp.: division

CONSOLIDA:
spp.: seed
C. ambigua: seed

CONVALLARIA spp.: division,
seed

CORDYLINE:
C. fruticosa: cane cuttings, air
layering
C. indivisa: cane cuttings

COREOPSIS:
spp.: division, seed, basal
cuttings
C. grandiflora: division, seed,
basal cuttings
C. rosea: division, seed, basal
cuttings

CORNUS:
spp.: suckers, hardwood
cuttings
C. alba: softwood cuttings,
French layering
C. florida: greenwood cuttings
C. sericea: softwood cuttings,
French layering

CORREA spp.: seed

CORYLOPSIS spp.: softwood
cuttings, layering

CORYLUS spp.: layering, stooling

COTINUS:
spp.: softwood cuttings,
layering
C. coggygria: French layering

COTONEASTER:

spp.: layering, semiripe
cuttings, seed, greenwood
cuttings
C. franchetii: layering, semiripe
cuttings, seed
C. horizontalis: layering,
semiripe cuttings, seed
C. salicifolius: layering,
semiripe cuttings, seed

CRASSULA:
C. argentea: upright leaf
cuttings
C. ovata: leaf petiole cuttings

CRATAEGUS spp.: grafting

CRINUM spp.: offsets

CROCUS spp.: division

CRYPTANTHUS spp.: seed

CUPRESSOCYPARIS *leylandii:* semiripe cuttings, hardwood
cuttings

CUPRESSUS spp.: hardwood
cuttings, grafting

CYCAS spp.: offsets

CYDONIA *oblonga:* hardwood
cuttings, stooling, French
layering, grafting

CYPRIPEDIUM spp.: division

DABOECIA *cantabrica:* dropping

DAHLIA spp. & cvs.: division, basal
cuttings, seed

DAPHNE spp.: softwood cuttings,
layering

DELPHINIUM:
spp.: seed, division, basal
cuttings
D. cardinale: seed, division,
basal cuttings
D. elatum: seed

DENDROBIUM spp.: seed

DEUTZIA:
spp.: greenwood cuttings,
hardwood cuttings, stooling
D. gracilis: softwood cuttings

DIANTHUS:
spp.: seed, layering
D. barbatus: seed, layering
D. deltoides: seed, layering

DICENTRA:
spp.: seed, division, semiripe
cuttings, root cuttings
D. eximia: seed, division
D. spectabilis: seed

DICTAMNUS spp.: root cuttings

DIEFFENBACHIA cvs.: cane cuttings,
air layering

DIGITALIS *purpureus:* seed

DIONAEA *muscipula:* leaf vein
cuttings

DRACAENA *fragrans:* air layering

DRACOCEPHALUM spp.: seed

DROSERA spp.: leaf vein cuttings

DYCKIA spp.: offsets

ECHINACEA:
spp.: seed, division, root
cuttings
E. purpurea: seed, division,
root cuttings

ECHINOCACTUS spp.: grafting

ECHINOCEREUS spp.: seed, grafting

ECHINOPS:
spp.: offsets
E. ritro: seed, root cuttings

ECHINOPSIS spp.: grafting

EPIPREMNUM *aureum:* seed,
layering

EPIPHYLLUM spp.: leaf petiole cuttings

EPITHELANTHA spp.: grafting

ERANTHIS:
spp.: division, seed
E. pinnatifida: division, seed

EREMURUS spp.: division

ERICA spp.: semiripe cuttings,
layering, dropping

ERYNGIUM:
spp.: seed, division, root
cuttings
E. bourgatii: seed, division, root
cuttings
E. giganteum: seed, division,
root cuttings

ERYTHRONIUM spp.: division, seed

ESCALLONIA spp.: hardwood
cuttings

ESCOBARIA spp.: grafting

EUCALYPTUS:
spp.: seed
E. cinerea: seed
E. citriodora: seed
E. gunnii: seed

EUCOMIS spp.: monocot leaf
cuttings

EUONYMUS:
spp.: layering, semiripe
cuttings, greenwood
cuttings
E. alatus: layering, semiripe
cuttings, greenwood
cuttings, softwood cuttings
E. fortunei: layering, semiripe
cuttings, greenwood
cuttings, softwood cuttings
E. japonicus: layering,
semiripe cuttings,
greenwood cuttings

EUPHORBIA:
spp.: seed, greenwood cuttings
E. griffithii: seed

EUSTOMA *grandiflorum:* seed,

division

FEROCACTUS spp.: seed
FIBIGIA clypeata: seed
FICUS:
  spp.: air layering, semiripe
    cuttings, seed, greenwood
    cuttings, hardwood cuttings,
    root cuttings
  F. benjamina: air layering,
    semiripe cuttings, seed
  F. carica: air layering, semiripe
    cuttings, seed
  F. elastica: air layering,
    semiripe cuttings, seed
FITTONIA spp: seed
FORSYTHIA spp.: layering,
  suckers, greenwood cuttings,
  softwood cuttings, semiripe
  cuttings, hardwood cuttings
FOTHERGILLA:
  spp.: layering
  F. major: softwood cuttings
FRAGARIA: offsets
FRANKLINIA alatamaha:
  softwood cuttings, layering,
  seed, hardwood cuttings
FREESIA spp.: division
FRITILLARIA spp.: division
FUCHSIA spp.: seed, greenwood
  cuttings

GAILLARDIA:
  spp.: seed, division, root
    cuttings
  G. aristata: seed
  G. grandiflora: seed, division,
    root cuttings
GALANTHUS:
  spp.: division, seed
  G. nivalis: division, seed
GALIUM spp.: division, seed
GARDENIA spp.: softwood
  cuttings, grafting, air layering
GARRYA spp.: hardwood cuttings
GAULTHERIA shallon: suckers
GENTIANA:
  spp.: seed, division, basal
    cuttings
  G. alba: seed, division, basal
    cuttings
  G. dendrologii: seed, division,
    basal cuttings
  G. lutea: seed, division, basal
    cuttings
  G. verna: seed, division, basal
    cuttings
GERANIUM:
  spp.: division, seed, softwood
    cuttings, hardwood cuttings

G. sanguineum: seed
  var. striatum: division, seed,
    softwood cuttings
GERBERA spp.: seed, division,
  basal cuttings
GLADIOLUS:
  spp.: division, seed
  G. callianthus: divisions, seed
  G. x hortulanus cvs: division
  G. tristis: division, seed
GLORIOSA spp.: division
GOMPHRENA globosa: seed
GONIOLIMON tataricum: seed
GREVILLEA spp.: seed, semiripe
  cuttings
GUZMANIA spp.: seed
GYPSOPHILA spp.: seed

HAMAMELIS spp.: layering, air
  layering, softwood cuttings,
  grafting
HARRISIA spp.: grafting
HAWORTHIA spp.: offsets
HEBE spp.: semiripe cuttings
HEDERA:
  spp.: layering, semiripe
    cuttings, hardwood cuttings,
    softwood cuttings
  H. helix: layering, semiripe
    cuttings, hardwood cuttings
  H. nepalensis: layering,
    semiripe cuttings,
    hardwood cuttings
HELENIUM autumnale: seed
HELIOPSIS helianthiodes: seed
HELLEBORUS:
  spp.: division, seed
  H. argutifolius: seed
  H. cyclophyllus: division, seed
  H. foetidus: seed
  H. x hybridus: division, seed
  H. niger: division, seed
  H. orientalis: division, seed
HEMEROCALLIS spp.: division,
  seed
HESPERIS matronalis: seed
HEUCHERA:
  cvs: division, seed
  spp: division, seed
HIBISCUS:
  spp.: layering, seed, softwood
    cuttings, semiripe cuttings,
    greenwood cuttings air
    layering, grafting
  H. moscheutos: layering, seed,
    softwood cuttings,
    semiripe cuttings
  H. rosa-sinensis: layering, seed,
    softwood cuttings,
    semiripe cuttings

H. syriacus: layering, seed,
  softwood cuttings, semi-
  ripe cuttings, hardwood
  cuttings
HIPPEASTRUM spp.: division
HOLODISCUS discolor:
  hardwood cuttings
HOSTA spp.: division, seed
HOYA spp.: leaf petiole cuttings,
  layering
HUMULUS spp.: root cuttings,
  layering
HYACINTHOIDES spp.: division
HYACINTHUS orientalis: division,
  scoring
HYDRANGEA:
  spp.: softwood cuttings,
    semiripe cuttings,
    hardwood cuttings
  H. anomala sbsp. petiolaris:
    layering
  H. arborescens: suckers
HYLOCEREUS spp.: grafting

IBERIS amara: seed
ILEX spp.: layering, softwood
  cuttings, semiripe cuttings,
  seed, greenwood cuttings, air
  layering, grafting
IMPATIENS walleriana: seed
INDIGOFERA spp.: semiripe
  cuttings
IPOMOEA:
  spp.: seed, softwood cuttings
  I. alba: seed, softwood cuttings
  I. x multifida: seed, softwood
    cuttings
  I. tricolor: seed, softwood
    cuttings
IRIS spp.: division, seed
ITEA virginica: softwood cuttings

JASMINUM:
  spp.: layering, semiripe
    cuttings, hardwood cuttings
  J. nudiflorum: layering,
    semiripe cuttings
  J. officinale: layering, semiripe
    cuttings
JUGLANS:
  spp.: French layering
  J. cinerea: grafting
  J. nigra: grafting
  J. regia: grafting
JUNIPERUS:
  spp.: layering, grafting,
    semiripe cuttings,
    hardwood cuttings
  J. chinensis: layering, grafting,
    semiripe cuttings,

hardwood cuttings
  J. communis: layering, grafting,
    semiripe cuttings,
    hardwood cuttings
  J. horizontalis: layering, grafting,
    semiripe cuttings,
    hardwood cuttings
  J. procumbens: layering,
    grafting, semiripe cuttings,
    hardwood cuttings

KALANCHOE:
  spp.: plantlets, division, seed,
    softwood cuttings, upright
    leaf cuttings
  K. blossfeldiana: division, seed,
    softwood cuttings
  K. daigremontiana: plantlets,
    seed
KALMIA spp.: greenwood
  cuttings, semiripe cuttings,
  layering
KERRIA japonica: suckers,
  greenwood cuttings, softwood
  cuttings, hardwood cuttings
KNIPHOFIA:
  spp.: seed, division
  K. caulescens: seed, division
KOELREUTERIA spp.: root
  cuttings
KOLKWITZIA amabilis:
  greenwood cuttings, softwood
  cuttings

LABURNUM spp.: seed, grafting:
  hardwood cuttings
LACHENALIA spp.: monocot leaf
  cuttings
LAGERSTROEMIA indica:
  softwood cuttings
LANTANA spp.: greenwood
  cuttings
LARIX spp.: hardwood cuttings
LATHYRUS:
  spp.: seed, division
  L. odoratus: seed, division
  L. sylvestris: seed, division
  L. vernus: seed, division
LAURUS spp.: semiripe cuttings
LAVANDULA:
  spp.: seed, layering, semiripe
    cuttings, stooling
  L. angustifolia: seed, layering,
    semiripe cuttings
  L. stoechas: seed, layering,
    semiripe cuttings
LAVATERA spp.: greenwood
  cuttings
LEMAIREOCEREUS spp.: seed
LEPTOSPERMUM spp.: seed

LEUCOTHOE spp.: greenwood cuttings

LEWISIA:
spp.: seed, offsets
L. brachycalyx: seed
L. cotyledon: seed, offsets
L. rediviva: seed

LEYCESTERIA formosa: hardwood cuttings

LIATRIS scariosa: seed

LIGULARIA:
spp.: seed, division, basal stem cuttings, root cuttings
L. dentata: seed, division, basal stem cuttings

LIGUSTRUM:
spp.: hardwood cuttings, layering
L. japonicum: softwood cuttings

LILIUM:
spp.: division, offsets, seed,
L. lancifolium: division, bulbils, seed

LIMONIUM:
spp.: seed, division, root cuttings
L. latifolium: seed, division
L. sinuatum: seed
L. tetragonum: seed

LIQUIDAMBAR spp.: layering

LOBELIA:
spp.: seed
L. cardinalis: seed

LONICERA:
spp.: layering, softwood cuttings, semiripe cuttings, hardwood cuttings
L. x americana: layering, softwood cuttings, semiripe cuttings
L. x brownii: layering, softwood cuttings, semiripe cuttings
L. x purpusii: softwood cuttings, semiripe cuttings
L. xylosteum: softwood cuttings, semiripe cuttings

LUNARIA annua: seed

LUPINUS:
spp.: seed, basal cuttings, division
L. polyphyllus: seed

LYCORIS spp.: division

MACHAERANTHERA tortifolia (syn. Xylorhiza): seed

MACLURA pomifera: root cuttings

MAGNOLIA spp.: greenwood cuttings, softwood cuttings, semiripe cuttings, layering,

air layering, budding

MALUS spp.: grafting, seed, stooling, French layering, budding

MAMMILLARIA spp.: grafting

MANGIFERA indica: grafting

MARANTA leuconeura: upright leaf cuttings

MIMULUS: seed

MOLUCCELLA laevis: seed

MONARDA spp.: seed, division, softwood cuttings

MONSTERA spp.: layering

MUSA spp.: offsets

MUSCARI spp.: division

MYOSOTIS spp: seed

MYRICA spp.: suckers, layering

NANDINA domestica: semiripe cuttings, seed, greenwood cuttings

NARCISSUS spp.: division

NELUMBO:
spp.: seed, division
N. lutea: seed, division
N. nucifera: seed, division

NEOREGELIA spp.: seed

NERIUM oleander: air layering

NYMPHAEA spp.: division, seed

OENOTHERA:
spp.: division, root cuttings
O. macrocarpa: seed

ONCIDIUM spp.: seed

ONOCLEA spp.: division, spores

OPUNTIA spp.: seed, grafting

ORIGANUM:
spp.: seed, division
O. majorana: seed, division
O. vulgare: seed, division

ORNITHOGALUM spp.: division

ORTEGOCACTUS macdougallii: offsets

ORTHOPHYTUM spp.: offsets

OSTEOSPERMUM spp.: greenwood cuttings

OXALIS:
spp.: seed, division
O. regnellii: seed, division
O. tetraphylla: seed, division

PACHYSANDRA:
spp.: division, seed, suckers
P. terminalis: division, seed

PAEONIA:
cvs: seed
spp.: seed, division, semiripe cuttings, root cuttings

PAPAVER:
spp.: seed, root cuttings

P. croceum: seed, root cuttings
P. orientale: seed, root cuttings
P. rhoeas: seed

PARTHENOCISSUS spp.: hardwood cuttings, layering

PAULOWNIA tomentosa: root cuttings

PELARGONIUM cvs.: greenwood cuttings

PENSTEMON:
spp.: seed, division, softwood cuttings, semiripe cuttings
P. caespitosus: seed, division, softwood cuttings, semiripe cuttings
P. digitalis: seed, division, softwood cuttingss, semiripe cuttings
P. pinifolius: seed, division, softwood cuttings, semiripe cuttings

PEPEROMIA spp.: leaf petiole cuttings

PERESKIA:
spp.: seed, softwood cuttings, semiripe cuttings
P. grandifolia: seed, softwood cuttings, semiripe cuttings

PERSEA americana: grafting

PETUNIA hybrida: seed

PHALAENOPSIS spp. & cvs: offsets, plantlets, softwood cuttings, seed

PHILADELPHUS:
spp.: softwood cuttings, semiripe cuttings, hardwood cuttings, stooling
P. coronarius: greenwood cuttings, softwood cuttings

PHILODENDRON spp.: layering

PHILOTHECA spp. (formerly Eriostemon): seed

PHLOX:
spp. seed
P. paniculata: division, root cuttings

PHOTINIA davidiana: softwood cuttings

PHYLLOSTACHYS:
spp.: division
P. aurea: division
P. aureosulcata var. aureocaulis: division
P. nigra: division

PHYSALIS:
spp.: root cuttings
P. alkekengi: seed

PHYSOSTEGIA spp.: seed, division

PICEA spp.: hardwood cuttings,

grafting

PIERIS:
spp.: seed, softwood cuttings, layering
P. japonica: seed, softwood cuttings, semiripe cuttings

PINUS:
spp.: seed, grafting, hardwood cuttings
P. strobus: seed, grafting
P. sylvestris: seed, grafting

PITTOSPORUM spp.: layering

PLATYCODON grandiflorus: seed, division

PODOCARPUS spp.: semiripe cuttings

PODOPHYLLUM spp.: division

POLEMONIUM borealis: seed

POLIANTHES tuberosa: division

POLYGONATUM spp.: division

POLYPODIUM spp.: division

POLYSTICHUM spp.: division

POPULUS:
spp.: suckers
P. balsamifera: greenwood cuttings

PORTULACA spp. seed

POTENTILLA spp.: greenwood cuttings, softwood cuttings, semiripe cuttings

PRIMULA:
spp.: seed, division, root cuttings
P. japonica: seed, division, root cuttings
P. marginata: seed
P. sinensis: seed, division, root cuttings

PRUNUS:
spp.: grafting, seed, greenwood cuttings, softwood cuttings, hardwood cuttings, French layering, budding
P. avium: grafting, seed, greenwood cuttings, softwood cuttings
P. cerasifera: grafting, seed, greenwood cuttings, softwood cuttings
P. maackii: grafting, seed, greenwood cuttings, softwood cuttings
P. serrula: grafting, seed, greenwood cuttings, softwood cuttings
P. virginiana: grafting, seed, greenwood cuttings, softwood cuttings

PSEUDOTSUGA menziesii:

hardwood cuttings

PYRACANTHA spp.: semiripe cuttings

PYRUS spp.: French layering, grafting, budding

QUERCUS:
ssp.: seed, grafting, semiripe cuttings
Q. agrifolia: seed, grafting, semiripe cuttings
Q. laurifolia: seed, grafting, semiripe cuttings
Q. virginiana: seed, grafting, semiripe cuttings

RANUNCULUS spp.: division

RHIPSALIS spp.: seed

RHODODENDRON:
spp.: layering, seed, softwood cuttings, grafting, semiripe cuttings, hardwood cuttings, air layering
R. arborescens: layering, seed, softwood cuttings, grafting

RIBES spp.: layering, stooling

ROSA:
cvs: seed
spp.: softwood cuttings, hardwood cuttings, layering, seed, root cuttings, budding
R. rugosa: budding, softwood cuttings, hardwood cuttings, seed, suckers

ROSMARINUS:
spp: seed, layering, semiripe cuttings
R. officinalis: seed, semiripe cuttings

RUBUS:
spp.: layering, stooling
R. fruticosus: tip layering
R. idaeus: tip layering

RUDBECKIA:
spp.: division
R hirta: seed

SAINTPAULIA:
cvs.: leaf petiole cuttings
hybrids: upright leaf cuttings
spp.: seed, division, leaf petiole cuttings

SALIX:
spp.: softwood cuttings, hardwood cuttings, grafting, French layering
S. alba: softwood cuttings, hardwood cuttings, grafting
S. babylonica var. *pekinensis*: softwood cuttings, hardwood cuttings, grafting

SALPIGLOSSIS *sinuata*: seed

SALVIA spp.: seed

SAMBUCUS spp.: suckers, softwood cuttings

SANGUISORBA *canadensis*: seed

SANSEVIERIA:
spp.: suckers, leaf cuttings
S. trifasciata: suckers, monocot leaf cuttings

SANTOLINA spp.: hardwood cuttings

SAXIFRAGA *stolonifera*: offsets

SCABIOSA spp.: seed

SCHEFFLERA spp.: layering, air layering

SCHISANDRA *chinensis*: semiripe cuttings

SCHIZANTHUS x *wisetonensis*: seed

SCHIZOPHRAGMA *hydrangeoides*: semiripe cuttings

SCILLA spp.: division

SEDUM:
spp.: division, seed, softwood cuttings, leaf petiole cuttings
S. morganianum: division, seed, softwood cuttings, leaf petiole cuttings

SELENICEREUS spp.: seed, grafting

SEMPERVIVUM spp.: seed, offsets

SILYBUM *marianum*: root cuttings

SINNINGIA spp.: leaf vein cuttings

SKIMMIA spp.: semiripe cuttings

SMITHIANTHA spp.: leaf vein cuttings

SOLANUM *lycopersicum*: grafting

SOLENOSTEMON spp.: seed, softwood cuttings

SORBARIA spp.: layering

SPARAXIS spp.: division

SPIRAEA spp.: softwood cuttings, semiripe cuttings, stooling

STACHYS *byzantina*: division

STACHYURUS *praecox*: layering

STEPHANANDRA spp.: hardwood cuttings, layering

STEPHANOTIS:
spp.: seed, softwood cuttings, layering, semiripe cuttings
S. floribunda: seed, softwood cuttings, layering, semiripe cuttings

STEWARTIA *pseudocamellia*: softwood cuttings

STOKESIA *laevis*: root cuttings

STREPTOCARPUS spp.: upright leaf cuttings

STYRAX spp.: layering

SYMPHYTUM spp.: root cuttings

SYRINGA:
spp.: suckers, layering, greenwood cuttings, seed, softwood cuttings, semiripe cuttings, root cuttings
S. x *chinensis*: layering, greenwood cuttings, seed
S. patula: layering, greenwood cuttings, seed
S. x *persica*: layering, greenwood cuttings, seed
S. x *prestoniae*: layering, greenwood cuttings, seed
S. vulgaris: suckers, layering, greenwood cuttings, seed

TANACETUM *parthenium*: seed

TAXUS spp.: semiripe cuttings, hardwood cuttings

TEUCRIUM spp.: greenwood cuttings

THUJA spp.: hardwood cuttings, grafting

TIBOUCHINA spp.: layering

TILLANDSIA spp: offsets

TOLMIEA *menziesii*: plantlets, division, seed

TRACHELOSPERMUM *jasminoides*: layering

TRICHOCEREUS spp.: seed

TRILLIUM spp.: division

TROLLIUS *europaeus*: seed

TSUGA spp.: hardwood cuttings, grafting

TULIPA spp.: division

TURBINICARPUS spp.: grafting

VACCINIUM:
spp.: layering, dropping, grafting
V. macrocarpon: dropping

VANDA spp.: seed

VERBASCUM:
spp.: seed, division, semiripe cuttings, root cuttings
V. chaixii f. *album*: seed, division, semiripe cuttings, root cuttings

VERBENA spp. & hybrids: seed

VERONICA:
spp.: seed, division, softwood cuttings
V. spicata: seed, division, softwood cuttings

VIBURNUM:
spp.: seed, greenwood cuttings, semiripe cuttings, hardwood cuttings, layering, French layering
V. x *burkwoodii*: softwood

cuttings
V. carlesii: seed, greenwood cuttings, semiripe cuttings, softwood cuttings
V. macrocephalum: seed, greenwood cuttings, semiripe cuttings

VINCA spp.: layering

VIOLA:
spp.: seed
V. x *wittrockiana*: seed

VITIS spp.: semiripe cuttings, hardwood cuttings, layering, grafting

VRIESEA spp.: seed

WALDSTEINIA spp.: layering

WEIGELA spp.: softwood cuttings, semiripe cuttings, hardwood cuttings

WISTERIA:
spp.: layering, softwood cuttings, root cuttings
W. floribunda: layering, softwood cuttings, root cuttings
W. frutescens: layering, softwood cuttings, root cuttings
W. sinensis: layering, softwood cuttings, root cuttings

YUCCA:
spp.: seed, offsets
Y. filamentosa: seed

ZANTEDESCHIA spp.: division

Index

Credits

Quarto would like to thank and acknowledge Writtle College for providing photography assistance and to thank and acknowledge the following for supplying photographs reproduced in this book:

18, 19tl	Chaselink UK Ltd www.chaselinkuk.com
19tr	Garden Supply Direct Ltd www.gardensupplydirect.co.uk
19b, 39t	Eden Greenhouses Ltd www.eden-greenhouses.com
55t	Craig Knowles/dk/Alamy
61br	Patrick Johns/Corbis
73b	Nikhilesh Haval www.nikreations.co.uk
74r	Alan Watson/Forest Light www.treesforlife.org.uk
87tr	Douglas Peebles/Corbis
99b	Ed Young/Corbis
100l	George D. Lepp/Corbis
113t	Karl Weatherly/Corbis
116r	Guenter Rossenbach/zefa/Corbis